Zeitschrift für Angewandte Mathematik und Mechanik

ZAMM

Applied Mathematics and Mechanics

Founded by Richard von Mises in 1921

Akademie-Verlag • Berlin

Volume 65 · 1985
Number 1

ISSN 0044-2267 ZAMM · Z. angew. Math. Mech., Berlin **65** (1985) 1, 1–64 EVP 18,— M

The **aim and scope** of ZAMM is, in agreement with the intentions of its founder: to publish new results and information on applied mathematics (mainly on numerical mathematics and applications of analysis) as well as on theoretical and applied mechanics.
The journal is of interest for persons working in applied mathematics and mechanics as well as on mathematical or mechanical questions in other sciences as physics and technical sciences, for instance mechanical and civil engineering.

Terms of subscription for the journal

Orders can be sent

- in the GDR: to Postzeitungsvertrieb, or to Akademie-Verlag, DDR-1086 Berlin, Leipziger Str. 3—4, PF-Nr. 1233;
- in the other socialist countries: to a book-shop for foreign language literature or to the competent news-distributing agency;
- in the FRG and Berlin(West): to a book-shop or to the wholesale distributing agency Kunst und Wissen, Erich Bieber OHG, Wilhelmstr. 4—6, D-7000 Stuttgart 1;
- in the other Western European Countries: to Kunst und Wissen, Erich Bieber GmbH, Dufourstr. 51, CH-8008 Zürich;
- in other countries: to the international book- and journal-selling trade, to Buchexport, Volkseigener Außenhandelsbetrieb der DDR, DDR-7010 Leipzig, Postfach 160; or to the Akademie-Verlag, DDR-1086 Berlin, Leipziger Str. 3—4, PF-Nr. 1233.

Manuscripts (in grammatical English or German language, master and one copy, formulas typewritten) should be addressed to the editorial office of ZAMM, Institut für Mechanik der Akademie der Wissenschaften der DDR, DDR-1199 Berlin, Rudower Chaussee 5. The title in English language and for papers of more than 10 pages a summary of 5 to 10 lines in English and (if possible) German and Russian laungages have to be added. Detailed remarks for authors which should be considered strictly are published subsequent to the last annual table of contents.

ZEITSCHRIFT FÜR
ANGEWANDTE MATHEMATIK UND MECHANIK
Herausgeber und Chefredakteur: Prof. Dr. Günter Schmidt, Berlin. Verlag: Akademie-Verlag, DDR-1086 Berlin, Leipziger Straße 3—4; Fernruf: 2236221 oder 2236229; Telex-Nr.: 114420; Bank: Staatsbank der DDR, Berlin, Kto.-Nr.: 6886-26-20712. Anschrift der Redaktion: Institut für Mechanik der Akademie der Wissenschaften der DDR, DDR-1199 Berlin, Rudower Chaussee 5; Fernruf: 6743639 oder 6743643.

Veröffentlicht unter der Lizenznummer 1282 des Presseamtes beim Vorsitzenden des Ministerrates der Deutschen Demokratischen Republik. Gesamtherstellung: VEB Druckerei „Thomas Müntzer", 5820 Bad Langensalza. Erscheinungsweise: Die Zeitschrift für Angewandte Mathematik und Mechanik erscheint monatlich. Die 12 Hefte eines Jahres einschließlich Tagungshefte bilden einen Band. Bezugspreis je Band 396,— M zuzüglich Versandspesen. Bezugspreis je Heft 33,— M. Bestellnummer dieses Heftes: 1009/65/1.

AN (EDV) 35937

Zeitschrift für Angewandte Mathematik und Mechanik

ZAMM

Applied Mathematics and Mechanics

Volume 65 1985 Number 1

ZAMM · Z. Angew. Math. u. Mech. 65 (1985) 1, 1–11

BRUNN, P. O.

The Time Dependent Deformation of a Capsule in an Arbitrary Stokes Flow

Der Aufsatz behandelt die zeitabhängige Deformation einer kugelförmigen Kapsel, die sich in einer viskosen Flüssigkeit befindet. Die grundlegenden Annahmen sind, daß (a) die Strömung in den Flüssigkeitsgebieten den Stokesschen Gleichungen gehorcht, (b) die Deformation klein ist und (c) die Membran rein elastisch ist. Die Voraussetzung (b) erfordert, daß entweder die Zeitkonstante der elastischen Verformungen klein ist im Vergleich zu einer charakteristischen Strömungszeit (das bedeutet eine genügend steife Membran und/oder ein genügend schwaches Strömungsfeld) oder die Flüssigkeit im Innern der Kugel hochviskos ist. Wegen dieser Voraussetzung ist es möglich, in erster Näherung eine lineare elastische Konstitutivgleichung für die Membran zu verwenden. Die allgemeinen Ergebnisse werden mit den entsprechenden Resultaten für Tröpfchenoberflächenspannungen verglichen. Der Fall des Sinkens in einer ruhenden Flüssigkeit, eine frei schwebende Kapsel in einem linearen Strömungsfeld und stationäre Zustände für quadratische Strömungsfelder (besonders für eine zylindrische Poiseuille-Strömung) werden untersucht.

This paper is concerned with the time-dependent deformation of a spherical capsule suspended in a viscous fluid. The basic assumptions are that (a) the flow in the fluid regions is governed by the Stokes' equations, (b) the deformation is small, and (c) the membrane is purely elastic. Condition (b) requires either the elastic time scale to be small in comparison with a characteristic flow time (i.e., a sufficiently stiff membrane and/or a sufficiently weak flow field) or the inside fluid to be highly viscous. It also allows us to employ a linearly elastic constitutive equation for the membrane, as first approximation. The general results obtained are compared with corresponding results for surface tension droplets. The case of sedimentation in a quiescent fluid, a capsule freely suspended in a linear flow field, and steady-state results for quadratic flow fields (with emphasis on circular Poiseuille flow) are studied.

В статье изучается зависящая от времени деформация сферической капсулы, находящейся в вязкой жидкости. Основными предположениями являются: а) поток в областях жидкости подчиняется уравнениям Стокса, б) деформации малы и в) мембрана чисто упругая. Предположение б) требует, чтобы или время упругих деформаций было мало по сравнению с характерным временем потока (это означает, что мембрана достаточно жёсткая и/или что поле потока достаточно слабое) или жидкость внутри сферы имеет большую вязкость. Благодаря этому предположению оказывается возможным применить для мембраны в первом приближении материальное уравнение линейной упругости. Общие результаты сравниваются с соответствующими результатами для поверхности напряжений капли. Исследуются случаи погружения в покоящейся жидкости, капсулы во взвешенном состоянии в линейном поле потока и стационарные состояния для квадратичного поля потока (особенно для циркуляционного потока Пуазейля).

1. Introduction

The behaviour of small particles in a viscous fluid at low (particle) Reynolds number is of great interest to both scientists and engineers [1]. If the particles are rigid, they are fixed in shape, and the mathematical problem is strictly linear. On the other hand, if a deformable particle is subjected to viscous forces, it will in general change its shape. Since this change in shape is determined by the details of the flow field and since these details depend upon the instantaneous geometry of the particle, the problem becomes highly nonlinear and approximations are necessary to analytically attack the problem.

One common approximation (which we shall rely on) assumes the particle to be initially spherical (radius R) and the deformation to be small. In this case, the change in shape is treated as a small correction to the basic problem of the motion of a spherical particle. Particular cases which have been studied concern liquid droplets [2—4] and elastic solids [5, 6].

Relatively recent is the interest in nonhomogeneous particles (i.e., particles containing another internal phase). Examples are fluid particles encapsulated by a liquid shell [7] and solid particles occluded within the interior of a droplet [8]. The case of fluid particles encapsulated by an elastic shell of arbitrary thickness [9, 10] has the purely elastic solid as one limit and the elastic membrane as the other limit. We shall follow [11] and call the latter particles "capsules". Such particles may be thought of as simple models of red blood cells. Although it cannot be expected to reproduce exactly all the striking properties of red blood cells, the model does lead to a "tank-treading" motion of the membrane in shear flow [11]. Furthermore, by using a simple "cell" method to estimate the shear viscosity of a concentrated suspension of capsules, BRENNEN [12] obtained excellent agreement with experimental results for red cell suspensions.

The present paper is concerned with the small deformation of a capsule. It represents one of the possible extensions of the previously mentioned studies since (a) the (linearly) elastic membrane need not be incompressible (b) the external flow may be arbitrary, being restricted only by the requirement that it obey the creeping motion equations, and (c) the complete time-dependent evolution of shape is considered. We shall assume that the deformation will lead to an almost spherical particle. After obtaining the change in shape, the parameters which guarantee that the deformation is indeed small can be read off. The extremely general equations obtained in this way will then be applied to the case of sedimentation in a quiescent flow and to the case of a "capsule" freely suspended in a linear shear flow.

II. The Deformation of a Spherical Membrane Subject to an Arbitrary Load

Consider a droplet (viscosity η_i) surrounded by an elastic membrane that is immersed in an incompressible newtonian solvent (viscosity η_0). The membrane thickness h is assumed to be so thin that any variations of elastic stresses across the membrane can be neglected. Let $\boldsymbol{q}$ denote the load vector (i.e., the force acting on the membrane per unit area relative to a zero net total force):

$$\int_S \mathrm{d}S\boldsymbol{q} = 0\,, \tag{2.1}$$

where the integration extends over the membrane area S. Let $\boldsymbol{T}^e$ denote the stress resultants of the membrane (i.e., the elastic stresses averaged over the thickness h). The equilibrium condition for the membrane then reads:

$$\frac{\partial}{\partial \boldsymbol{r}_s}\cdot \boldsymbol{T}^e + \boldsymbol{q} = 0\,, \tag{2.2}$$

where $\partial/\partial \boldsymbol{r}_s$ is the surface gradient operator, $\partial/\partial \boldsymbol{r}_s = (\boldsymbol{\delta} - \hat{\boldsymbol{n}}\hat{\boldsymbol{n}})\cdot \partial/\partial \boldsymbol{r}$ and $\hat{\boldsymbol{n}}$ is the unit outer normal of S.

Now assume that the capsule, when unstressed, is a sphere of radius a (outer normal $\hat{\boldsymbol{N}}$). Then, if α — some measure of the magnitude of $\boldsymbol{q}$ — is small, S will deviate but slightly from its spherical equillibrium shape. Phrased differently, the deformation of the membrane will be small. Consequently, it suffices to concentrate on the limit of a linearly elastic membrane, as a first approximation. Furthermore, since $\hat{\boldsymbol{n}} = \hat{\boldsymbol{N}} + O(\alpha)$, the equilibrium condition (2.2) can then be applied at S_0, the undeformed spherical surface. The solution to this equation furnishes the displacement vector $\boldsymbol{u}$ and this, in turn, allows us to evaluate S.

To be more specific, let $\boldsymbol{\chi}$ denote any point on the middle surface of the stress-free membrane ($|\boldsymbol{\chi}| = R$). If $\boldsymbol{x}$ denotes the same point in the deformed state, then the displacement $\boldsymbol{u}$ is defined by:

$$\boldsymbol{x} = \boldsymbol{\chi} + \boldsymbol{u}\,. \tag{2.3}$$

The magnitude of $\boldsymbol{x}$ specifies the deformed shape of the membrane and is, to first order in α, (Note that $\boldsymbol{u}$ is of order α) given by:

$$r_s \equiv |\boldsymbol{x}| = R\left[1 + \frac{u_r}{R}\right], \qquad u_r = \boldsymbol{u}\cdot\hat{\boldsymbol{N}}\,. \tag{2.4}$$

The displacement field $\boldsymbol{u}$ leads to the strain:

$$\boldsymbol{\varepsilon}^* = \frac{1}{2}\left(\frac{\partial \boldsymbol{u}}{\partial \boldsymbol{\chi}} + \left(\frac{\partial \boldsymbol{u}}{\partial \boldsymbol{\chi}}\right)\right) \tag{2.5}$$

and we only need the two-dimensional version of $\boldsymbol{\varepsilon}^*$, termed $\boldsymbol{\varepsilon}$:

$$\boldsymbol{\varepsilon} = (\boldsymbol{\delta} - \hat{\boldsymbol{N}}\hat{\boldsymbol{N}})\cdot \boldsymbol{\varepsilon}^*\cdot (\boldsymbol{\delta} - \hat{\boldsymbol{N}}\hat{\boldsymbol{N}})\,. \tag{2.6}$$

As emphasized before, it is sufficient to concentrate on a linearly elastic membrane. Thus, if E denotes Young's modulus and ν denotes Poisson's ratio, the membrane stresses $\boldsymbol{T}^e$ are related to $\boldsymbol{\varepsilon}$ by Hooke's law:

$$\boldsymbol{T}^e = \frac{Eh}{1+\nu}\left[\boldsymbol{\varepsilon} + \frac{\nu}{1-\nu}(\boldsymbol{\delta} - \hat{\boldsymbol{N}}\hat{\boldsymbol{N}})\, I_1\right], \tag{2.7}$$

with I_1 the first invariant of $\boldsymbol{\varepsilon}$ (its trace).

Zeitschrift für Angewandte Mathematik und Mechanik

ZAMM

Applied Mathematics and Mechanics

Volume 64

vierundsechzigster Jahrgang 1984

AKADEMIE-VERLAG · BERLIN

Inhaltsübersicht

Originalarbeiten

(B = Bericht; Ber. = Berichtigung; H = Hauptaufsatz; HV = Hauptvortrag; KM = Kleine Mitteilung; N = Nachrichten; V = Vortragsauszug)
Der Buchstabe T vor der Seitenzahl weist darauf hin, daß der betreffende Beitrag in den Sonderheften 4 und 5 (GAMM-Tagung 1983) erschienen ist. Der Buchstabe M weist auf Beiträge aus dem Vortragsheft 10 (1. Mechanikkongreß der DDR 1983) hin.

Seite

Nachrichten etc.

Buchbesprechungen

Hinweise für Autoren

Manuskripte (zweifach, Original und eine Kopie) und andere den Inhalt der Zeitschrift betreffende Zuschriften sind dem Herausgeber der ZAMM entweder über ein Mitglied des Herausgebergremiums oder direkt über folgende Anschrift zuzuleiten:

Zeitschrift für Angewandte Mathematik und Mechanik
Institut für Mechanik
der Akademie der Wissenschaften der DDR
DDR-1199 Berlin, Rudower Chaussee 5

Bei Einreichen des Manuskripts hat der Autor verbindlich zu erklären, daß sein Aufsatz noch in keiner anderen allgemein zugänglichen Zeitschrift veröffentlicht worden ist bzw. kein Antrag auf Veröffentlichung in einer solchen Zeitschrift läuft.

Es wird erwartet, daß der Autor über eine weitere Kopie verfügt. Die Manuskripte müssen in ihrer äußeren Gestalt folgenden Anforderungen genügen, anderenfalls sie zurückgewiesen werden können oder in der Bearbeitung lange Verzögerungen eintreten:

1. Die Manuskripte sind in einwandfreier sprachlicher Form (einschließlich Orthographie, Interpunktion) einzureichen.

2. Die Manuskriptseiten sind nur einseitig zu beschreiben. Der Text ist mit Maschine mit weitem Zeilenabstand (zweizeilig) und mit mindestens 3 cm breitem Rand zu schreiben.

3. Die **Formeln** müssen einwandfrei lesbar mit Maschine oder in Druckschrift und mit großem Zeilenabstand geschrieben sein. Für mehrfach wiederkehrende umfangreiche oder drucktechnisch komplizierte Terme sind abkürzende Bezeichnungen einzuführen. Bei Exponentialausdrücken mit längerem Exponenten ist weitgehend von der Schreibweise exp(...) Gebrauch zu machen. Die verwendeten Formelzeichen sind in einer für den Setzer eindeutigen Weise zu kennzeichnen und gegebenenfalls in einer besonderen Liste zu erläutern. Um unnötige Korrekturen zu vermeiden, ist genau zu unterscheiden zwischen v, ν, γ und r; φ und Φ; Ψ und ψ; 0 (Null), O, o, Θ und θ; l, l und e; K, k und $\varkappa$; i und ι; ξ und ζ; x, X, $\times$ und χ; u und n. Griechische Buchstaben können rot unterstrichen werden. Die wichtigsten, insbesondere auch die im weiteren Verlauf des Textes zitierten Formeln sind am rechten Rand der Manuskriptseiten in runden Klammern durchzunumerieren.

4. Es ist darauf zu achten, daß alle benutzten Abkürzungen definiert sind. Es wird gebeten, Textabkürzungen wie Dgl., eq. zu vermeiden.
Sätze, Hilfssätze, Korollare, Definitionen, Beweise u. dgl. sind dadurch hervorzuheben, daß ihnen in Sperrdruck das betreffende Wort (Satz..., Beweis..., usw.) vorangestellt wird. Die Aussage eines Satzes oder Hilfssatzes wird einheitlich kursiv gedruckt. Das Ende einer solchen Aussage oder eines Beweises ist im Manuskript zu markieren. Bei Hauptaufsätzen ist im Interesse einer übersichtlicheren Gliederung und besseren Lesbarkeit weitgehend von der Möglichkeit Gebrauch zu machen, einzelne Passagen (Beweise, Bemerkungen, ...) durch Kleindruck gegenüber dem Haupttext zurücktreten zu lassen und die Stellen durch den Hinweis „Petit" zu kennzeichnen.

5. Als Vorlagen für die **Abbildungen** sind saubere Zeichnungen in Bleistift oder Tusche (Beschriftung nur mit Bleistift) auf gesonderten Blättern und in einem genügend großen, eine Verkleinerung gestattenden Format erforderlich. Da Korrekturen in den Abbildungen schwierig sind, ist hier besonders sorgfältig zwischen ähnlichen Buchstaben zu unterscheiden (vgl. 3).

6. Auf **Literatur** ist im Text durch Nummern in eckigen Klammern zu verweisen. Das nach diesen Nummern geordnete Literaturverzeichnis erscheint am Ende des Aufsatzes. Es wird gebeten, darin die Literatur nach folgendem Muster anzuführen:

bei Büchern:
Householder, A. S., The Theory of Matrices in Numerical Analysis, Blaisdell Publ., Comp., New York/Toronto/London 1964, p. 74

bei Zeitschriftenartikeln:
Weissinger, J., Zur nichtlinearen Theorie der ungleichförmigen Umströmung von Profilen, Z. Angew. Math. u. Mech. **50** (1970), 337–346.

bei Sammelwerken:
Barber, J. R., Thermoelastic contact problems. In: de Pater, A. D.; Kalker, J. J. (ed.), The Mechanics of the Contact Between Deformable Bodies. Proc. Symp. IUTAM, Enschede 20. bis 23. 8. 74. Delft: University Press 1975, pp. 177–205.

Russischsprachige Zitate können in kyrillischen Buchstaben oder in einer der gebräuchlichen Transliterationen geschrieben werden.
Handelt es sich bei der zitierten Literaturstelle um eine Übersetzung, so ist anzugeben, wo bzw. in welcher Zeitschrift, in welchem Jahr und in welcher Sprache das Original erschienen ist.

7. Den Manuskripten, die als Hauptaufsätze bestimmt sind, ist auf gesondertem Blatt eine **Zusammenfassung** von 5–10 Zeilen in englischer, deutscher und russischer Sprache beizufügen. Falls sie nicht in deutscher oder russischer Sprache geliefert werden kann, ist wenigstens eine Liste der vorkommenden Fachausdrücke in der fehlenden Sprache erwünscht.

8. Dem Manuskript ist die **Anschrift der Autoren** (mit akademischem Titel und Institut) beizufügen, die am Ende des Aufsatzes zur Information des Lesers veröffentlicht wird. Darüber hinaus sind diejenigen Anschriften anzugeben, an welche die Korrekturabzüge und die Sonderdrucke zu senden sind. Ändert sich während der Laufzeit einer Arbeit eine Anschrift, so ist dies zur Vermeidung von Fehlleitungen und langen Verzögerungen dem Herausgeber der ZAMM mitzuteilen. Die Autoren erhalten von den Hauptaufsätzen 75, von den Kleinen Mitteilungen 25 Sonderdrucke ohne Berechnung, darüber hinaus weitere gegen Berechnung.

Remarks for authors

Manuscripts (master and one copy) and other correspondence relating to the contents of the journal should be addressed to the editor of ZAMM, either through any one member of the editorial board or direct to the following address:

Zeitschrift für Angewandte Mathematik und Mechanik
Institut für Mechanik
der Akademie der Wissenschaften der DDR
DDR-1199 Berlin, Rudower Chaussee 5

On submitting manuscripts authors must give the explicit assurance that the paper submitted has not previously been published in, nor been submitted for publication to, another public journal.

Authors should have a copy of their manuscript at their own. The following rules in the set-up of manuscripts should be observed to avoid the risk of delayed publication or even rejection.

1. The manuscripts should be well written (including spelling and punctuation).

2. The text should be typewritten using only one side of the sheet, with adequate line spacing (2-line spacing) and a margin of not less than 3 cm.

3. **Formulas** must be clearly legible (typewritten or written clearly) and with adequate line spacing. Long terms that occur especially frequently or terms that present special difficulty to type-setters should be abbreviated, if possible. In exponential expressions with comparatively long exponents the abbreviation "exp(...)" should preferably be used throughout. Signs used in formulas should be clearly recognizable to type-setters or, if necessary, explained in an appended list. To avoid tiresome corrections special care should be taken to clearly distinguish between ν, γ, v and r; φ and Φ; ψ and Ψ; 0(zero), O, o, Θ and θ; l, l and e; K, k and $\varkappa$; i and ι; ξ and ζ; x, X, $\times$ and χ u and n. It would be helpful if Greek letters are underlined in red. Major formulas, in particular the ones quoted throughout the text, should be numbered continually on the right margin (in parentheses).

4. All abbreviations must be defined, and abbreviated words or text sections should be avoided.
Theorems, corollaries, definitions, proofs, etc. should be emphasized by printing, in spaced type, the relevant term (theorem..., proof..., etc.) in front of the expression, which, in its turn, will be printed uniformly in italics. The end of a theorem, proof. etc. should be marked in the manuscript. Certain passages (proof, remarks, etc.) in articles and monographs may be printed in smaller type to ensure easy readability and should, therefore, be marked "petit" in the manuscript.

5. **Illustrations** should be in the form of clear pencil or ink drawings (captions, legends, etc. in pencil only!) on separate sheets, big enough to be scaled down.
Since corrections in illustrations are difficult special care must be taken to differentiate clearly between similar letters (see paragraph 3).

6. **References** should be numbered in brackets throughout the text, with the bibliography according to these numbers appended. References should follow the model printed below:

books:
HOUSEHOLDER, A. S., The Theory of Matrices in Numerical Analysis, Blaisdell Publ. Comp., New York/Toronto/London 1964, p. 74.

articles:
WEISSINGER, J., Zur nichtlinearen Theorie der ungleichförmigen Umströmung von Profilen, Z. Angew. Math. u. Mech. **50** (1970), 337–346.

proceedings etc.:
BARBER, J. R., Thermoelastic contact problems. In: DE PATER, A. D.; KALKER, J. J. (ed.), The Mechanics of the Contact Between Deformable Bodies. Proc. Symp. IUTAM, Enschede 20.–23. 8. 74. Delft: Delft University Press 1975, pp. 177–205.

Russian quotations can be written in Cyrillic letters or in customary transcriptions.
In cases of translated references, the original source (name of periodical, year of publication, original language) should be stated.

7. A concise **summary** (5 to 10 lines) in English, German or Russian should be provided on a separate sheet for manuscripts intended as major articles. If a summary cannot be provided in German or Russian, a list of the technical terms of the missing language will be appreciated.

8. Authors should state their academic title, institute and address which will be published at the end of the contribution for the information of readers. Addresses should also be stated of persons to whom galley proofs and reprints are to be sent. The editors of ZAMM should be notified of any change of the author's address occurring after submission of a contribution, so as to avoid any postal errors and delays. Authors of major articles will be supplied with 75 free reprints, authors of other contributions will receive 25. All additional reprints have to be paid for.

ZEITSCHRIFT FÜR ANGEWANDTE MATHEMATIK UND MECHANIK
Herausgeber und Chefredakteur: Prof. Dr. Günter Schmidt, Redaktion: Dipl.-Math. F. Dudel, Dr. W. Heinrich, B. Müllenhaupt, Dr. H. Weinert, Institut für Mechanik der Akademie der Wissenschaften der DDR. Verlag: Akademie-Verlag, DDR-1086 Berlin, Leipziger Straße 3–4; Fernruf: 2236229 oder 2236221; Telex-Nr.: 114420; Bank: Staatsbank der DDR, Berlin, Kto.-Nr.: 6836-26-20712. Anschrift der Redaktion: Institut für Mechanik der Akademie der Wissenschaften der DDR, DDR-1199 Berlin, Rudower Chaussee 5; Fernruf: 6743639 u. 6743643.

Veröffentlicht unter der Lizenznummer 1282 des Presseamtes beim Vorsitzenden des Ministerrates der Deutschen Demokratischen Republik. Gesamtherstellung: VEB Druckerei „Thomas Müntzer", DDR-5820 Bad Langensalza. Erscheinungsweise: Die Zeitschrift für Angewandte Mathematik und Mechanik erscheint monatlich. Die 12 Hefte eines Jahres einschließlich Tagungshefte bilden einen Band. Bezugspreis 1985 je Band 396,– M zuzüglich Versandspesen
Bezugspreis je Heft 33,– M

AN (EDV) 35937

Having specified the problem, it remains to evaluate the displacement under the action of an arbitrary but given load $\boldsymbol{q}$. To treat a problem of such generality, we have to establish certain notational conventions. Let $\boldsymbol{A}^l$ denote an arbitrary tensor of order l and $\overline{\boldsymbol{A}^l}$ the symmetric irreducible part of $\boldsymbol{A}^l$. For tensors which by definition are symmetric and irreducible, we shall omit the $\overline{\cdots}$ notation. For example, $\hat{\boldsymbol{r}}^l$ denotes the l-folded dyadic product of the unit vector, $\hat{\boldsymbol{r}} = 1/R\boldsymbol{\chi}$. In defining:

$$\boldsymbol{Y}^l \equiv (2l-1)!!\,\overline{\hat{\boldsymbol{r}}^l} \tag{2.8}$$

the multiple potentials $r^{-(l+1)}\boldsymbol{Y}^l$ are automatically characterized as symmetric and irreducible. Finally, the notation $\boldsymbol{A}^l(\cdot k)\,\boldsymbol{B}^m$ stands for a k-folded scalar product (nesting convention implied) between the tensors $\boldsymbol{A}^l$ and $\boldsymbol{B}^m$, $l \geqq k, m \geqq k \geqq 0$ (i.e., a tensor of order $l + m - 2k$). We also will have to use the operator $\Delta^{(l)}$, a tensor of order $2l$ uniquely defined by its projection property:

$$\overline{\boldsymbol{A}^l} = \Delta^{(l)}(\cdot l)\,\boldsymbol{A}^l\,. \tag{2.9}$$

Now, let us assume that $\boldsymbol{q}$ can be expanded as:

$$\boldsymbol{q} = \sum_{l=0} \boldsymbol{q}^{l+1}(\cdot l)\,\boldsymbol{Y}^l\,. \tag{2.10}$$

(Note that in spherical polar coordinates, r, θ, φ, the expansion $f(\boldsymbol{r}) = \sum_{l=0} \boldsymbol{f}^l(\cdot l)\,\boldsymbol{Y}^l$ is equivalent to $f(\boldsymbol{r}) = \sum_{n,m} f_{nm}(r) \times \times Y_n^m(\theta, \varphi)$, where the functions Y_n^m are called surface spherical harmonics). Since the expansion tensor $\boldsymbol{q}^{l+1}$ is by definition symmetric and irreducible with respect to its last l indices, it can uniquely be decomposed into its irreducible parts [13]:

$$\boldsymbol{q}^{l+1} = -\boldsymbol{e}\colon (\boldsymbol{q}^{l;1}(l-1)\,\Delta^{(l)}) + \Delta^{(l)}\cdot(l-1)\,\boldsymbol{q}^{l-1;2} + \boldsymbol{q}^{l+1;3}\,. \tag{2.11}$$

Here, $\boldsymbol{e}$ is the alternator and the tensors $\boldsymbol{q}^{l;i}$, $i = 1, 2, 3$ are the (three) symmetric and irreducible parts of $\boldsymbol{q}^{l+1}$. Note that $\boldsymbol{q}^{l;1}$ and $\boldsymbol{q}^{l;3}$ are defined only for $l \geqq 1$, while $\boldsymbol{q}^{l;2}$ is defined for all $l \geqq 0$.

In order for (2.1) to be satisfied, we need $\boldsymbol{q}^1 = 0$. This implies the restriction:

$$\boldsymbol{q}^{1;3} = 0 \tag{2.12}$$

so that the summation of eq. (2.10) starts from $l = 1$.

With $\boldsymbol{q}$ given and the membrane rheology specified (eq. 2.7), the equilibrium condition (2.2) at $r = R$ can be solved for $\boldsymbol{u}$. We expand this solution as:

$$\boldsymbol{u} = \sum_{l=0} \boldsymbol{U}^{l+1}(\cdot l)\,\boldsymbol{Y}^l \tag{2.13}$$

and denote by $\boldsymbol{U}^{l;1}$, $\boldsymbol{U}^{l-1;2}$ $\boldsymbol{U}^{l+1;3}$ the three symmetric and irreducible parts of the expansion tensor $\boldsymbol{u}^{l+1}$ (cf. eq. 2.11). Then, u_r is given by:

$$u_r = \sum_{l=0}\left[(l+1)\,\boldsymbol{U}^{l;2} + \frac{1}{2l-1}\,\boldsymbol{U}^{l;3}\right](\cdot l)\,\boldsymbol{Y}^l\,. \tag{2.14}$$

If the volume enclosed by the membrane cannot change and since we assumed the center of the coordinate system to be at the center of the particle, the summation in eq. (2.14) has to start from $l = 2$. Consequently, the following must be true:

$$U^{0;2} = 0\,, \qquad 2\boldsymbol{U}^{1;2} + \boldsymbol{U}^{1;3} = 0\,. \tag{2.15a), (2.15b}$$

Relying heavily on the properties of irreducible tensors and of $\boldsymbol{Y}^l$ (details can be found in the appendices of two previous papers [13, 14], it may be shown that eq. (2.2) implies:

$$\boldsymbol{q}^{l;1} = \frac{\eta_0}{2(1+\nu)\,R\tau_0}(l-1)(l+2)\,\boldsymbol{U}^{l;1}\,, \tag{2.16a}$$

$$\boldsymbol{q}^{l;2} = \frac{\eta_0}{(1-\nu^2)\,R\tau_0}\,\frac{l+2}{2l+1}\left\{[(l^2+l+1)+\nu(2l+1)]\,\boldsymbol{U}^{l;2} - \frac{l-1}{2l-1}\,\boldsymbol{U}^{l;3}\right\}, \tag{2.16b}$$

$$\boldsymbol{q}^{l;3} = \frac{\eta_0}{(1-\nu^2)\,R\tau_0}\,\frac{l-1}{2l+1}\{-l(l+1)(l+2)(2l-1)\,\boldsymbol{U}^{l;2} + [(l^2+l+1)-\nu(2l+1)]\,\boldsymbol{U}^{l;3}\}\,, \tag{2.16c}$$

with:

$$\tau_0 = \frac{R\eta_0}{Eh} \tag{2.17}$$

a characteristic time.

While (2.12) is automatically satisfied, the restriction (2.15a) requires:

$$q^{0;2} = 0 \tag{2.18}$$

To get the physical meaning of that result, consider a no-flow (equilibrium) situation in which the (constant) pressure in the solvent is p_∞ while $\hat{p}$ is the inside pressure. In this case, $q^{0;2} = \hat{p} - p_\infty$ and (2.18) requires the equality of inside and outside pressure. Since p_∞ is immaterial, we set it zero, i.e.:

$$\hat{p} = p_\infty = 0 \tag{2.19}$$

This contrasts the behavior of a surface tension droplet, where an excess inside pressure $\hat{p}$ is needed to balance the surface tension σ. However, for a surface tension droplet the stress resultant is $\sigma(\boldsymbol{\delta} - \hat{\boldsymbol{n}}\hat{\boldsymbol{n}})$, irrespective of the type of flow. Thus, in this case the interface is always in the same constant stress state. In our case, the stress state changes with the applied load, and since we assumed the membrane to be stress-free in the no-flow equilibrium situation, eq. (2.19) could have been anticipated from the beginning.

III. The Capsule in an Arbitrary Stokes' Flow Field

Before proceeding to consider the capsule in an arbitrary Stokes' flow field, it is necessary to utilize one other piece of information which can be extracted from the results of the last chapter. This concerns the term containing the (pseudo) vector $\boldsymbol{U}^{1;1}$, which describes a rigid body rotation of the membrane. By (2.16a), there should be no loading in that case (i.e., $\boldsymbol{q}^{1;1} = 0$). Thus, it seems natural to subtract this solid body rotation out of the problem. If $\boldsymbol{\omega}$ denotes the angular velocity of the undeformed membrane, the choice of a rotating reference configuration $\boldsymbol{\chi}$:

$$\dot{\boldsymbol{\chi}} = \boldsymbol{\omega} \times \boldsymbol{\chi} \tag{3.1}$$

will obviously do the trick. We call this (rotating) frame of reference body frame. Relative to fixed $\boldsymbol{\chi}$, we then have:

$$\dot{\boldsymbol{U}}^{1;1} = 0 \, . \tag{3.2}$$

We note in passing that a flow, which is steady in the non-rotating frame, will be time-dependent in the body frame. As long as inertia effects can safely be neglected, this causes no problems: the linear Stokes' equations are valid in either frame of reference. Unless stated otherwise, we shall use the body frame and shall denote by $\boldsymbol{r}$ the position vector in that frame (i.e., a vector colinear with $\boldsymbol{\chi}$ and of magnitude r, the distance from the particle center).

Let $(\boldsymbol{V}(\boldsymbol{r}, t), \boldsymbol{P}(\boldsymbol{r}, t))$ denote an arbitrary but given velocity-pressure field (We shall work solely with the hydrodynamic pressure.) in the Stokes regime that the fluid would have if the particle were absent. This being the case, there are at most three independent symmetric and irreducible parts of the velocity gradient of any order [15]. Explicitly, these are the for l-th order gradient of $\boldsymbol{V}$ (a tensor of order $l + 1$).

$$\boldsymbol{B}^{l+1} = \frac{1}{(l+1)!\,(2l+1)!!} \overline{\frac{\partial^l}{\partial \boldsymbol{r}^l} \boldsymbol{V}} \, , \tag{3.3a}$$

$$\boldsymbol{A}^l = \frac{1}{(l+1)!\,(2l-1)!!} \Delta^{(l)}(\cdot l) \overline{\frac{\partial^l}{\partial \boldsymbol{r}^l}} \times \boldsymbol{V} \, , \tag{3.3b}$$

and:

$$\boldsymbol{C}^{l-1} = \eta_0{}^2 \boldsymbol{B}^{l-1} \, . \tag{3.3c}$$

Note that $6\boldsymbol{B}^2$ is the (undisturbed) rate of strain dyadic $\dot{\boldsymbol{E}}$, and $\boldsymbol{A}^1$ is half the vorticity vector.

As before, we expand:

$$\boldsymbol{V}(\boldsymbol{r}) = \sum_{l=0} \boldsymbol{V}^{l+1}(\cdot l)\, \boldsymbol{Y}^l \, , \qquad \boldsymbol{P}(\boldsymbol{r}) = \sum_{l=0} \boldsymbol{P}^l(\cdot l)\, \boldsymbol{Y}^l \, , \tag{3.4a), (3.4b}$$

where the expansion tensors $\boldsymbol{V}^{l+1}$ and $\boldsymbol{p}^l$ depend upon r and t. Denoting by $\boldsymbol{V}^{l;i}$, $i = 1, 2, 3$, the irreducible parts of $\boldsymbol{V}^{l+1}$ (cf. eq. 2.11), it is rather straightforward to show:

$$\boldsymbol{P}^l = r^l \boldsymbol{C}^l \, , \qquad \boldsymbol{V}^{l;1} = l r^l \boldsymbol{A}^l \, , \tag{3.5a), (3.5b}$$

$$\boldsymbol{V}^{l;2} = - \frac{l r^{l+1}}{\eta_0 (l+1)\,(2l+1)\,(2l+3)} \boldsymbol{C}^l \, , \tag{3.5c}$$

$$\boldsymbol{V}^{l;3} = l(2l-1) \left[r^{l-1} \boldsymbol{B}^l + \frac{r^{l+1}}{2(2l+1)\,\eta_0} \boldsymbol{C}^l \right] . \tag{3.5d}$$

Note that in accordance with (2.19), we put $C^0 = 0$.

The disturbance fields $(\boldsymbol{v}, p)$ produced in the suspending medium may be expanded analogously to (3.4). Expressing the corresponding expansion tensors as:

$$\boldsymbol{p}^l = \frac{1}{r^{l+1}} \boldsymbol{c}^l \, , \qquad \boldsymbol{v}^{l;1} = \frac{l}{r^{l+1}} \boldsymbol{a}^l \, , \tag{3.6a), (3.6b}$$

$$\boldsymbol{v}^{l;2} = \frac{l}{r^{l+2}} \boldsymbol{b}^l + \frac{r^{-l}}{2(2l+1)\,\eta_0} \boldsymbol{c}^l \, , \qquad \boldsymbol{v}^{l;3} = \frac{l+1}{(2l+1)\,\eta_0 r^l} \boldsymbol{c}^l \, , \tag{3.6c), (3.6d}$$

guarantees that the Stokes' equations are satisfied [13]. The symmetric and irreducible tensors $\boldsymbol{a}^l, \boldsymbol{b}^l$, and $\boldsymbol{c}^l$ are unknown. They can be expressed as surface moments of the stress and velocity fields [15]. It thus turns out that $-8\pi\eta_0 \boldsymbol{a}^1$ is the hydrodynamic couple and $-4\pi \boldsymbol{c}^1$ is the hydrodynamic force $\boldsymbol{F}^h$ exerted upon the particle. If we admit gravitational forces, we have:

$$\boldsymbol{a}^1 = 0 \, , \qquad \boldsymbol{c}^1 = - \frac{R^3}{3} \Delta\varrho \boldsymbol{g} \, , \tag{3.7a), (3.7b}$$

where. $\Delta\varrho$ is the density difference between the suspending fluid and the inside fluid. (We have neglected the gravitational force of the membrane).

Finally, the inside fields ($\mathfrak{V}$, P) are expanded in the form of eq. (3.4). The Stokes' equations are satisfied if we put (Again, by eq. (2.19) we set $\boldsymbol{C}^0 = 0$)

$$\boldsymbol{P}^l = r^l \boldsymbol{C}^l \,, \quad \mathfrak{V}^{l;1} = l r^l \boldsymbol{A}^l \,, \tag{3.8a), (3.8b}$$

$$\mathfrak{V}^{l;1} = -\frac{l r^{l+1}}{(l+1)(2l+1)(2l+3)\eta_i}\boldsymbol{C}^l \,, \quad \mathfrak{V}^{l;3} = l(2l-1)\left[r^{l-1}\boldsymbol{B}^l + \frac{r^{l+1}}{2(2l+1)\eta_i}\boldsymbol{C}^l\right]. \tag{3.8 c, d}$$

The symmetric and irreducible tensors $\boldsymbol{A}^l$, $\boldsymbol{B}^l$, and C^l are unknown but are related to the "outside" tensors by employing the continuity of the velocity field across the membrane. Via the independence of the irreducible tensors this implies [13]:

$$\mathfrak{V}^{l;i} = \boldsymbol{V}^{l;i} + \boldsymbol{v}^{l;i} \,, \qquad i = 1, 2, 3 \,. \tag{3.9}$$

For simplicity we introduce:

$$\tilde{\boldsymbol{B}}^l = \boldsymbol{B}^l + \frac{R^2}{2(2l+1)\eta_0}\boldsymbol{C}^l \,, \qquad \tilde{\boldsymbol{B}}^l = \boldsymbol{B}^l + \frac{R^2}{2(2l+1)\eta_i}\boldsymbol{C}^l \,, \qquad \tilde{\boldsymbol{b}}^l = \boldsymbol{b}^l + \frac{R^2}{2(2l+1)\eta_0}\boldsymbol{c}^l \,, \tag{3.10 a, b, c}$$

to obtain by eq. (3.9):

$$\boldsymbol{A}^l = \boldsymbol{A}^l + \frac{1}{R^{l+1}}\boldsymbol{a}^l \,, \tag{3.11a}$$

$$\frac{1}{\eta_i}\boldsymbol{C}^l = \frac{1}{\eta_0}\mathrm{C}^l - \frac{(l+1)(2l+1)(2l+3)}{l R^{2l+3}}\tilde{\boldsymbol{b}}^l \,, \tag{3.11b}$$

$$\boldsymbol{B}^l = \tilde{\boldsymbol{B}}^l + \frac{l+1}{l(2l-1)(2l+1)\eta_0 R^{2l-1}}\boldsymbol{c}^l \,. \tag{3.11c}$$

Note that we have, by eq. (2.8a):

$$\mathbf{A}^1 = \boldsymbol{A}^1 \,, \tag{3.12}$$

showing that the inside rotates with the same angular velocity as the outside fluid.

Having established the general form of the velocity-pressure fields in the respective regions, we also know the stress tensors $\boldsymbol{T}^0$ (outside) and $\boldsymbol{T}^i$ (inside). This, then, allows us to evaluate the load $\boldsymbol{q}$:

$$\boldsymbol{q} = \hat{\boldsymbol{n}} \cdot (\boldsymbol{T}^0 - \boldsymbol{T}^i) - \frac{1}{S}\boldsymbol{F}^h \tag{3.13}$$

Although $\boldsymbol{q}$ is defined for the membrane surface S we can, under the restrictions laid down in the last chapter, calculate $\boldsymbol{q}$ at $r = R$ as a first approximation. The result is of the form (2.10) with the expansion tensors:

$$\boldsymbol{q}^{l;1} = l(l-1)\,R^{l-1}(\eta_0 - \eta_i)\,\boldsymbol{A}^l - l[(l+2)\,\eta_0 + (l-1)\,\eta_i]\,R^{-(l+2)}\boldsymbol{a}^l \,, \tag{3.14a}$$

$$\boldsymbol{q}^{l;2} = \frac{2l^2+4l+3}{(l+1)(2l+1)(2l+3)}\left[-1 + \frac{\eta_i}{\eta_0}\right]R^l\boldsymbol{C}^l - \frac{2l(l+2)\,\eta_0 + (2l^2+4l+3)\,\eta_i}{l R^{l+3}}\tilde{\boldsymbol{b}}^1 + \frac{1}{R^2}\boldsymbol{c}^1\delta_{l1} \,, \quad l \geqq 1 \tag{3.14b}$$

$$\boldsymbol{q}^{l;3} = 2(l-1)\,l(2l-1)\,(\eta_0 - \eta_i)\,R^{l-2}\tilde{\boldsymbol{B}}^l + \frac{1}{R^2}\boldsymbol{c}\,\delta_{l1} - \frac{(2l^2+1)\,\eta_0 + 2(l^2-1)\,\eta_i}{(2l+1)\,\eta_0 R^{l+1}}\boldsymbol{c}^l \,. \tag{3.14c}$$

Note that eqs. (2.12) and (2.18) are satisfied. Utilizing these results in the previously established identities (2.16) leads, for $l = 1$, to the relation:

$$\frac{3\eta_0}{(1-\nu)\,R\tau_0}\boldsymbol{U}^{1;2} = -\frac{3}{10}R\left[1 - \frac{\eta_i}{\eta_0}\right]\mathrm{C}^1 - \frac{6\eta_0}{R^4}\left[1 + \frac{3}{2}\frac{\eta_i}{\eta_0}\right]\tilde{\boldsymbol{b}}^1 + \frac{1}{R^2}\boldsymbol{c}^1 \,. \tag{3.15}$$

For $l > 1$, all coefficients appearing in (2.16) are nonzero. This then allows us to invert the equations, i.e., to obtain $\boldsymbol{a}^l$, $\tilde{\boldsymbol{b}}^l$, and $\boldsymbol{c}^l$ in terms of $\boldsymbol{U}^{l;1}$, $\boldsymbol{U}^{l;2}$, and $\boldsymbol{U}^{l;3}$ (note that our main interest concerns the deformation). The result is:

$$\boldsymbol{a}^l = R^{2l+1}d_{1l}\boldsymbol{A}^l + \frac{R^{l+1}}{\tau_0}d_{2l}\boldsymbol{U}^{l;1} \,, \tag{3.16a}$$

$$\tilde{\boldsymbol{b}}^l = \frac{R^{2l+3}}{\eta_0}e_{1l}\boldsymbol{C}^l + \frac{R^{l+2}}{\tau_0}e_{2l}\boldsymbol{U}^{l;2} + \frac{R^{l+2}}{\tau_0}e_{3l}\boldsymbol{U}^{l;3} \,, \tag{3.16b}$$

$$\boldsymbol{c}^l = R^{2l-1}\eta_0 f_{1l}\tilde{\boldsymbol{B}}^l + \frac{R^l\eta_0}{\tau_0}f_{2l}\boldsymbol{U}^{l;2} + \frac{R^l\eta_0}{\tau_0}f_{3l}\boldsymbol{U}^{l;3} \,. \tag{3.16c}$$

With:

$$\varkappa = \eta_i/\eta_0 \tag{3.17}$$

the inside viscosity relative to the fluid and:

$$d_l = -l(l+2)\left[1 + \frac{l-1}{l+2}\varkappa\right], \tag{3.18a}$$

$$e_l = -\left[2(l+2) + \frac{2l^2+4l+3}{l}\varkappa\right], \qquad f_l = -\frac{2l^2+1+2(l^2-1)\varkappa}{2l+1}, \qquad \text{(3.18b), (3.18c)}$$

the coefficients are:

$$d_{1l} = \frac{l(l-1)(1-\varkappa)}{d_l}, \qquad d_{2l} = \frac{(l-1)(l+2)}{2(1+\nu)\,d_l}, \qquad \text{(3.19a), (3.19b)}$$

$$e_{1l} = \frac{2l^2+4l+3}{(l+1)(2l+1)(2l+3)\,e_l}(1-\varkappa), \tag{3.20a}$$

$$e_{2l} = \frac{l+2}{(2l+1)(1-\nu^2)\,e_l}[(l^2+l+1)+\nu(2l+1)], \tag{3.20b}$$

$$e_{3l} = -\frac{(l-1)(l+2)}{(1-\nu^2)\,e_l(2l+1)(2l-1)}, \qquad f_{1l} = \frac{2(l-1)\,l(2l-1)}{f_l}(1-\varkappa), \qquad \text{(3.20c), (3.21a)}$$

$$f_{2l} = \frac{(l-1)(l+2)(l+2)(2l-1)\,l(l+1)}{(1-\nu^2)(2l+1)f_l}, \tag{3.21b}$$

$$f_{3l} = \frac{(l-1)[(l^2+l+1)-\nu(2l+1)]}{(1-\nu^2)(2l+1)f_l}, \tag{3.21c}$$

Having satisfied the equilibrium condition (2.3), it remains to invoke the continuity of fluid and membrane velocity, i.e.:

$$\dot{\boldsymbol{x}} = \boldsymbol{V} + \boldsymbol{v}, \quad \text{at} \quad r = R. \tag{3.22}$$

By utilizing the results obtained so far, we thus get for all $l \neq 0$ (Note that only the tensor $U^{0;2}$ is defined for $l = 0$. But $U^{0;2}$ has been shown to be zero):

$$\dot{\boldsymbol{U}}^{l;1} = -\frac{l^2(2l+1)\,R^l}{d_l}\boldsymbol{A}^l + \frac{l}{\tau_0}d_{2l}\boldsymbol{U}^{l;1}, \tag{3.23a}$$

$$\dot{\boldsymbol{U}}^{l;2} = \frac{R^{l+1}}{(l+1)\,\eta_0 e_l}\boldsymbol{C}^l + \frac{e_{2l}}{\tau_0}\boldsymbol{U}^{l;2} + \frac{e_{3l}}{\tau_0}\boldsymbol{U}^{l;3} - \frac{1}{R\eta_0 e_l}\boldsymbol{c}^1\,\delta_{1l} \tag{3.23b}$$

$$\dot{\boldsymbol{U}}^{l;3} = -\frac{R^{l-1}l(2l-1)^2}{f_l}\tilde{\boldsymbol{B}}^l + \frac{2}{3\eta_0 R}\boldsymbol{c}^1\delta_{l1} + \frac{(l+1)\,f_{2l}}{(2l+1)\,\tau_0}\boldsymbol{U}^{l;2} + \frac{(l+1)\,f_{3l}}{(2l+1)\,\tau_0}\boldsymbol{U}^{l;3}. \tag{3.23c}$$

In these equations, $\dot{\boldsymbol{U}}^{l;i}$ denotes the time rate of change of the tensors $\boldsymbol{U}^{l;i}$ with χ fixed. Equation (2.42a) implies $\dot{\boldsymbol{U}}^{1;1} = R\boldsymbol{A}^1$. This complies with (3.2) only if $\boldsymbol{A}^1$, the angular velocity of the fluid relative to $\boldsymbol{\omega}$, vanishes. Thus, if $\boldsymbol{\Omega}$ denotes the angular velocity of the fluid relative to a fixed frame, then $\boldsymbol{\omega} = \boldsymbol{\Omega}$, as expected.

Note also that in order for eq. (2.15b) to be satisfied, the following must be true:

$$\boldsymbol{U}^{1;2} = -\frac{1}{2}\boldsymbol{U}^{1;3} = \tau_0(1-\nu)\left[\left(1+\frac{3}{2}\varkappa\right)\tilde{\boldsymbol{B}}^1 - \frac{R^2}{6\eta_0}\boldsymbol{C}^1 + \frac{1+\varkappa}{\eta_0 R}\boldsymbol{c}^1\right]. \tag{3.24}$$

The details of this somewhat exceptional case, which does not lead to any deformation, will be investigated in Chapter 4.

Equations (3.23 a—c) are the equations governing the time evolution of the tensors $\boldsymbol{U}^{l;i}$ in the body frame. Since they are linear, these equations are readily integrated. The result becomes especially simple if we assume that the flow was started at time $t = 0$:

$$\boldsymbol{U}^{l;1} = \frac{l^2(2l+1)\,R^l}{d^l}\int_0^t \mathrm{d}t^1 \exp\left(\frac{t-t^1}{\tau_{l*}}\right)\boldsymbol{A}^l, \qquad \tau_{l*} = -\frac{\tau_0}{l d_{2l}}, \qquad l \geqq 2 \tag{3.25a}$$

$$\boldsymbol{U}^{l;2}(t) = \int_0^t \mathrm{d}t^1 g_{1l}(t-t^1)\,\boldsymbol{C}^l(t^1) + \int_0^t \mathrm{d}t^1 g_{2l}(t-t^1)\,\tilde{\boldsymbol{B}}^l(t^1), \tag{3.25b}$$

$$\boldsymbol{U}^{l;3}(t) = \int_0^t \mathrm{d}t^1 h_{1l}(t-t^1)\,\boldsymbol{C}^l(t^1) + \int_0^t \mathrm{d}t^1 h_{2l}(t-t^1)\,\boldsymbol{B}^l(t^1), \qquad l \geqq 2. \tag{3.25c}$$

In these equations, the memory functions g_{il}, h_{il}, $i = 1, 2$, are the inverse Laplace transforms of:

$$\overline{g_{1l}} = \frac{\tau_0 R^{l+1}\left[T - \frac{l+1}{2l+1}f_{3l}\right]}{\eta_0(l+1)\,e_l\Delta_l}, \qquad \overline{g_{2l}} = -\tau_0 R^{l-1}\frac{l(2l-1)^2}{f_l\Delta_l}e_{3l}, \qquad \text{(3.26a), (3.26b)}$$

$$\overline{h_{1l}} = \frac{\tau_0 R^{l+1}}{\eta_0}\frac{f_{2l}}{(2l+1)\,e_l\Delta_l}, \qquad \overline{h_{2l}} = -\tau_0 R^{l-1}\frac{l(2l-1)^2(T-e_{2l})}{f_l\Delta_l}, \qquad \text{(3.26c), (3.26d)}$$

with:

$$\Delta_l = T^2 - T\left[\frac{l+1}{2l+1} f_{3l} + e_{2l}\right] + \frac{(l^2-1)(l+2)}{(2l+1)(1-\nu^2)\, e_l f_l}, \qquad T = \tau_0 s\,. \tag{3.27), (3.28}$$

The actual time-dependent behaviour of the memory functions thus depends crucially upon the nature of the two roots $T_{1/2}$ of Δ_l. Since for general l, $\varkappa$, and ν, this is a rather formidable task, we shall investigate these roots in detail only for one specific example in Chapter 4.

In order to transform eqs. (3.23) or (3.26) in to the nonrotating frame, the (possibly time-dependent) angular velocity $\boldsymbol{\omega} = \boldsymbol{\Omega}$ has to be specified. Assuming this to be the case, let us emphasize which frame is being used by the use of a caret. Thus, if $\boldsymbol{S}$ denotes a particular tensor in the body frame, its equivalent in the nonrotating frame would be designated as $\check{\boldsymbol{S}}$. For the careted tensors so defined, the equations (3.23) hold, provided the time-derivative appearing on the lefthand side is interpreted as corotational or Jaumann derivative D/Dt. Realizing that the results obtained so far are actually first order results allows us to replace D/Dt by a partial time derivative in the nonrotating (but cotranslating) frame. This being the case, eqs. (3.25) can directly be used for the tensors marked by a caret.

For the special case in which the undisturbed flow far from the particle is constant in time (relative to the nonrotating frame), it is then easy to employ the final value theorem for Laplace transforms to obtain the steady state solutions (for $t \to \infty$).

$$\check{\boldsymbol{U}}^{l;1} = R^l \tau_0 \frac{2(2l+1)}{(l-1)(l+2)} (1+\nu) \check{\boldsymbol{A}}^l\,, \qquad l \geqq 2 \tag{3.29a}$$

$$\check{\boldsymbol{U}}^{l;2}(\infty) = -\frac{R^{l+1}\tau_0}{\eta_0} \frac{[(l^2+l+1) - \nu(2l+1)]}{(l+1)(l+2)(2l+1)} \check{\boldsymbol{C}}^l + \frac{l(2l-1)}{l+1} R^{l-1}\tau_0 \check{\check{\boldsymbol{B}}}^l \tag{3.29b}$$

$$\check{\boldsymbol{U}}^{l;3}(\infty) = -\frac{R^{l+1}\tau_0}{\eta_0} \frac{l(2l-1)}{2l+1} \check{\boldsymbol{C}}^l + \frac{l(2l-1)^2}{(l^2-1)} [l^2+l+1) + \nu(2l+1)] R^{l-1}\tau_0 \check{\check{\boldsymbol{B}}}^l \tag{3.29c}$$

The steady-state shape of the microcapsule in an arbitrary Stokes' field flow is thus given by eq. (2.4) with:

$$u_r = \sum_{l=2} R^{l-1}\tau_0 \left[\frac{l(2l-1)(2l+1)}{(l-1)(l+1)} (l+\nu) \check{\check{\boldsymbol{B}}}^l - \frac{R^2}{\eta_0}\frac{(l+1-\nu)}{l+2} \check{\boldsymbol{C}}^l\right] (\cdot l)\, \check{\boldsymbol{Y}}^l\,. \tag{3.30}$$

For example, for a neutrally buoyant microcapsule in a homogeneous flow, we have $l = 2$ with $6\check{\boldsymbol{B}}^2 = \boldsymbol{E}$, where $\boldsymbol{E}$ is the rate of strain dyadic. The shape of the membrane thus becomes:

$$r = R[1 + \tfrac{5}{3}(2+\nu)\tau_0 \boldsymbol{E} : \check{\boldsymbol{Y}}^2]\,. \tag{3.31}$$

This implies that the particle deforms into an ellipsoid whose principal axis coincides with the principal axes of $\boldsymbol{E}$. Equation (2.50) agrees with published results [16].

On the other hand, for a quadradic flow, we have to add the contribution from $\check{\check{\boldsymbol{B}}}^3 = \check{\boldsymbol{B}}^3$ (Note that in a quadradic flow, the vector $\boldsymbol{C}^1$ appears, which does not contribute to the deformation; it is handled in the $l = 1$ terms). Thus, the term:

$$\tfrac{105}{8} R^2 \tau_0 (3+\nu)\, \check{\check{\boldsymbol{B}}}^3 (\cdot 3)\, \check{\check{\boldsymbol{Y}}}^3 \tag{3.32a}$$

has to be added to eq. (3.31). For example, for channel flow:

$$\boldsymbol{V} = (\alpha + \beta x_3 + \gamma x_3^2)\, \hat{\boldsymbol{i}}$$

we have:

$$\check{\boldsymbol{B}}^3 (\cdot 3)\, \check{\boldsymbol{Y}}^3 = \tfrac{1}{3}\gamma(\hat{x}_1 \hat{x}_3^2 - \tfrac{1}{5}\hat{x}_1) \tag{3.32b}$$

and this is exactly of the form as encountered in the case of droplets [17].

If we insert eqs. (3.29) into (3.16), we obtain for $\check{\boldsymbol{a}}^l$, $\check{\boldsymbol{b}}^l$, and $\check{\boldsymbol{c}}^l$ the same expressions as for a rigid sphere [13]. By eq. (3.30), the particle will be almost spherical whenever the elastic time constant τ_0 is small in comparison with some time, τ, characteristic of the flow. But τ_0 is small when Eh becomes large (i.e., when the membrane is nearly rigid). Consequently, this result could have been anticipated from the beginning. This kind of approach also makes it clear why the equilibrium solution is independent of the properties of the inside fluid (to first order).

Obviously, for this result to be valid, the coefficients on the righthand side of eq. (3.23) have to be nonzero. Thus, the case of a highly viscous capsule has to be excluded. In this limit, we immediately extract from eq. (3.24):

$$\check{\check{\boldsymbol{B}}}^1 = -\frac{2}{3\eta_0 R} \check{\boldsymbol{c}}^1 = \frac{1}{6\pi\eta_0 R} \boldsymbol{F}^h\,. \tag{3.33}$$

Recalling the definition of $\tilde{\boldsymbol{B}}^1$ and $\boldsymbol{B}^1$, we see that eq. (3.33) is nothing but the (classical) Faxén law for a rigid sphere:

$$\boldsymbol{F}^h = 6\pi\eta_0 R\left[\boldsymbol{V}(0) + \frac{1}{6\eta_0} \Delta^2\, \boldsymbol{V}(0)\right]. \tag{3.34}$$

Utilizing eq. (2.16), one sees that all coefficients appearing in these equations vanish asymptotically for $\varkappa \to \infty$. This implies:

$$\dot{U}^{l;i} = 0\,, \qquad i = 1, 2, 3\,, \quad l \text{ arbitrary}. \tag{3.35}$$

Thus, relative to the reference configuration $\boldsymbol{X}$, rotating with angular velocity $\boldsymbol{\omega} = \boldsymbol{\Omega}$, there is no displacement in that limit. For $\varkappa^{-1}$ small but nonzero, we can thus repeat the whole analysis as we now shall demonstrate.

IV. Highly Viscous Capsules

We now consider the case of highly viscous capsules, $\varkappa \to \infty$. Denoting by a subscript zero the $\varkappa^{-1} = 0$ limit of the appropriate quantity, we have from our previous analysis (recall $\boldsymbol{a}^1 = 0$):

$$\boldsymbol{a}_0^l = -R^{2l+1}\boldsymbol{A}^l\,, \qquad l > 1\,, \tag{4.1a}$$

$$\tilde{\boldsymbol{b}}_0^l = R^{2l+3}\frac{l}{(l+1)(2l+1)(2l+3)\eta_0}\boldsymbol{C}^l\,, \qquad \boldsymbol{c}_0^l = -\eta_0 R^{2l-1}\frac{l(2l-1)(2l+1)}{(l+1)}\tilde{\boldsymbol{B}}^l\,, \tag{4.1b), (4.1c}$$

and:

$$U_0^{l;i} = 0\,, \qquad i = 1, 2, 3\,. \tag{4.2}$$

We now expand for $\varkappa^{-1}$ small, but nonzero any tensor $\boldsymbol{S}$ as:

$$\boldsymbol{S} = \boldsymbol{S}_0 + \frac{1}{\varkappa}\boldsymbol{S}_1 + \dots\,, \tag{4.3}$$

($\boldsymbol{S}_0$ may be zero), and repeat the calculation. Putting:

$$\boldsymbol{X} = \boldsymbol{x} \tag{4.4}$$

as new (and old) reference configuration (recall 2.3), the displacement $\boldsymbol{u}$ has to be order $\varkappa^{-1}$. This implies, by eq. (2.16), that the loads $\boldsymbol{q}$ will also be of that order which allows us by eqs. (3.14) to evaluate $\boldsymbol{a}_1^l$, $\boldsymbol{b}_1^l$, $\boldsymbol{c}_1^l$. This result is:

$$\boldsymbol{a}_1^l = \begin{cases} 0\,, & l = 1\,, \qquad (4.5\text{a}) \\ \dfrac{2l+1}{l-1}R^{2l+1}\boldsymbol{A}^l\,, & l > 1\,, \qquad (4.5\text{b}) \end{cases}$$

$$\tilde{\boldsymbol{b}}_1^l = -\frac{R^{2l+3}l}{\eta_0(l+1)(2l^2+4l+3)}\boldsymbol{C}^l - \frac{1}{6}R^3\tilde{\boldsymbol{B}}^1\delta_{l1}\,, \tag{4.6}$$

$$\boldsymbol{c}_1^l = \begin{cases} \eta_0 R^{2l-1}\dfrac{l(2l-1)^2(2l+1)^2}{2(l-1)(l+1)^2}\tilde{\boldsymbol{B}}^l\,, & l > 1\,, \qquad (4.7\text{a}) \\ \dfrac{\eta_0 R}{2}\tilde{\boldsymbol{B}}^1 + \dfrac{R^3}{6}\boldsymbol{C}^1\,, & l = 1\,. \qquad (4.7\text{b}) \end{cases}$$

Note that for eq. (4.7b) we had to rely on eq. (3.34). Its consequences are:

$$\boldsymbol{F}^h = 6\pi\eta_0 R\left(1 - \frac{\varkappa^{-1}}{3}\right)\left\{\boldsymbol{V}(0) + \frac{R^2}{6}\left(1 - \frac{2}{3}\varkappa^{-1}\right)\Delta^2\boldsymbol{V}(0)\right\}, \tag{4.8}$$

which is Faxén's law for a highly viscous droplet [18].

The deformation is governed by the kinematic condition (3.32). By means of the results obtained, it reads:

$$\dot{U}_1^{l;1} = \frac{l}{R^{l+1}}\boldsymbol{a}_1^l\,, \qquad \dot{U}_1^{l;2} = \frac{1}{R^{l+2}}\tilde{\boldsymbol{b}}_1^l\,, \qquad \dot{U}_1^{l;3} = \frac{l+1}{(2l+1)\eta_0 R^l}\boldsymbol{c}_1^l\,. \tag{4.9 a, b, c}$$

The condition (2.15b) is identically satisfied. At this stage of the calculation, it is no longer possible to interpret the time derivatives taken relative to the rotating reference frame as ordinary partial time derivatives. The time evolution of the tensor characterizing the membrane shape is (see 2.14):

$$\frac{D}{Dt}\left[(l+1)\check{\boldsymbol{U}}^{l;2} + \frac{1}{2l-1}\check{\boldsymbol{U}}^{l;3}\right] = \frac{lR^{l-1}}{\varkappa}\left[\frac{(2l-1)(2l+1)}{2(l^2-1)}\check{\tilde{\boldsymbol{B}}}^l - \frac{R^2}{\eta_0[2l^2+4l+3]}\check{\boldsymbol{C}}^l\right]. \tag{4.10}$$

Thus, at this order the actual evolution of the shape of the particle is unaffected by the membrane properties altogether. For a homogeneous flow ($l = 2$), the righthand side becomes:

$$\frac{5}{6}\frac{R}{\varkappa}\boldsymbol{E} \tag{4.11a}$$

and is thus identical to the corresponding case of a highly viscous droplet. The same should be true for higher order nonhomogeneous flow fields. For example, for the third order tensors, the forcing term on the righthand side of (3.10) is:

$$\frac{105}{16}\frac{R^2}{\varkappa}\check{\boldsymbol{B}}^{(3)}\,. \tag{4.11b}$$

For a capsule moving along the axis in circular Poiseuille flow, the change in shape would be governed solely by that term.

V. Examples

V.1 Sedimentation

With P interpreted as hydrodynamic pressure, the only nonzero tensor characterizing the undisturbed fields is $\boldsymbol{B}^1$, the negative of the terminal settling velocity $\boldsymbol{V}_t$. Utilizing eqs. (3.15) and (3.34) furnishes the relation:

$$\boldsymbol{B}^1 + \frac{2}{3\eta_0 R}\boldsymbol{c}^1 + \frac{2}{R^3}\tilde{\boldsymbol{b}}^1 = 0 . \tag{5.1}$$

This implies that the normal (radial) component of the velocity is zero at the surface of the particle. (This result is generally valid for the $l = 1$ fields, i.e., even with $\boldsymbol{C}^1 \neq 0$). In conjunction with eqs. (2.14/15), we thus see that no deformation of the capsule is possible (i.e., it will remain a sphere in Stokes' flow). For the case of a surface tension droplet, this result is well known.

The quantity of main interest is $\boldsymbol{c}^1$, which is simply proportional to the hydrodynamic force. For highly viscous capsules, we have already shown:

$$\boldsymbol{F}^h = -6\pi\eta_0 R\left(1 - \frac{\varkappa^{-1}}{3}\right)\boldsymbol{V}_t . \tag{5.2}$$

For finite $\varkappa$ but $\tau_0(V_t/R) \ll 1$, we turn to eqs. (3.32) to obtain:

$$\boldsymbol{F}^h = -6\pi\eta_0 R\boldsymbol{V}_t , \tag{5.3}$$

i.e., Stokes' law. Thus, the drag force is totally unaffected by the membrane properties.

V.2 Homogeneous Flow

We now consider a microcapsule freely suspended ($\boldsymbol{c}^1 = 0$, $\boldsymbol{a}^1 = 0$) in shear flow. If G is some measure of the rate of strain dyadic, $\boldsymbol{E} = 6\check{\boldsymbol{B}}^2$ (e.g., $G = \sqrt{\boldsymbol{E}:\boldsymbol{E}}$), the characteristic time τ of the flow is G^{-1}. Since the case of highly viscous capsules is (to the order to which our results are valid) identical to the highly viscous droplet which has already received considerable attention [3, 4], we shall not consider it and will concentrate instead on the case in which the membrane is sufficiently stiff to guarantee $\tau_0/\tau \ll 1$. This being the case, we have by eq. (2.35):

$$\check{\boldsymbol{U}}^{2;2}(t) = \frac{1}{6}\int_0^t dt' g_{22}(t - t')\,\boldsymbol{E}(t') , \qquad \check{\boldsymbol{U}}^{2;3}(t) = \frac{1}{6}\int_0^t dt' h_{22}(t - t')\,\boldsymbol{E}(t') . \tag{5.4a), (5.4b}$$

The two roots, $T_{1/2}$ of Δ_2, as defined by eq. (2.37), turn out to be:

$$T_{1/2} = \frac{\pm\chi - \left[(7 + \nu) + \frac{\varkappa}{8}(49 - 3\nu)\right]}{12(1 - \nu^2)\left(1 + \frac{2}{3}\varkappa\right)\left(1 + \frac{19}{16}\varkappa\right)} , \tag{5.5}$$

with:

$$\chi^2 = (25 + 14\nu + 25\nu^2) + \varkappa\left(\frac{165}{4} + 7\nu + \frac{175}{4}\nu^2\right) + \varkappa^2\left(\frac{1185}{64} - \frac{147}{32}\nu + \frac{1225}{64}\nu^2\right). \tag{5.6}$$

Since the two roots are real, negative, and distinct, we get by defining:

$$\tau_i = -\frac{\tau_0}{T_i} , \qquad i = 1, 2 . \tag{5.7}$$

$$g_{22}(t) = \frac{2R}{\chi}\left[e^{-t/\tau_1} - e^{-t/\tau_2}\right] , \tag{5.8}$$

$$h_{22}(t) = \frac{5R}{\left(1 + \frac{2}{3}\varkappa\right)\chi}\left\{\left[\chi + \frac{7}{5} + 5\nu + \frac{7}{8}\varkappa\left(-\frac{3}{5} + 5\nu\right)\right]e^{-t/\tau_1} - \right.$$
$$\left. - \left[-\chi + \frac{7}{5} + 5\nu + \frac{7}{8}\varkappa\left(-\frac{3}{5} + 5\nu\right)\right]e^{-t/\tau_2}\right\}. \tag{5.9}$$

Note that for the case of an incompressible membrane, $\nu = 1/2$, the two characteristic times τ_1 and τ_2 are in agreement with [19]. Furthermore, they have also been obtained and their numerical values discussed by considering in detail an elastic shell of finite thickness h/R by taking $h/R \to 0$ [9].

The second order tensor, which governs the actual deformation of the capsule, becomes:

$$\left[3\check{\boldsymbol{U}}^{2;2} + \frac{1}{3}\check{\boldsymbol{U}}^{2;3}\right] = \frac{5R}{18\left(1 + \frac{2}{3}\varkappa\right)}\int_0^t dt'\boldsymbol{E}(t')\left[I_1 e^{-\frac{t-t'}{\tau_1}} - I_2 e^{-\frac{t-t'}{\tau_2}}\right] ,$$

with:

$$I_{1/2} = \pm 1 + [5(1 + \nu) + \tfrac{5}{8}\varkappa(3 + 7\nu)]\,\chi^{-1}\,. \tag{5.10}$$

Note that since the times τ_i are positive, the equilibrium limit of eq. (5.10) (given by eq. 2.41) is stable.

Having determined the displacement field $\boldsymbol{u}$, the tensors $\tilde{\boldsymbol{b}}^2$ and $\boldsymbol{c}^2$, characterizing the flow field, follow from eqs. (3.16b) and (3.16c). For example, $\check{\boldsymbol{c}}^2$ becomes:

$$\check{\boldsymbol{c}}^2 = \frac{10}{9} R^3\eta_0 \frac{1-\varkappa}{1+\frac{2}{3}\varkappa}\boldsymbol{E} + \frac{8\eta_0 R}{(1-\nu^2)\left(1+\frac{2}{3}\varkappa\right)\tau_0}\check{\boldsymbol{U}}^{2;2} - \frac{(7-5\nu)\,R^2\tau_0}{9(1-\nu^2)\left(1+\frac{2}{3}\varkappa\right)}\check{\boldsymbol{U}}^{2;3}\,. \tag{5.11}$$

This quantity is of interest in suspension rheology. For a dilute suspension of particles (number density n), the bulk stress $\langle \boldsymbol{T}\rangle$ is:

$$\langle \boldsymbol{T}\rangle + \text{const.}\,\boldsymbol{\delta} = 2\eta_0\langle \boldsymbol{E}\rangle - 4\pi n\check{\boldsymbol{c}}^2\,, \tag{5.12}$$

where $\langle \boldsymbol{E}\rangle$ is the bulk rate of strain dyadic. To first-order in the concentration, we can replace the rate of strain dyadic $\boldsymbol{E}$, appearing in the expression for $\check{\boldsymbol{c}}^2$, by $\langle \boldsymbol{E}\rangle$. We thus find that the suspension shows viscoelastic behaviour with two time constants. (For $\nu = 1/2$, this result has been reported before [9, 19]). If the equilibrium expressions for $\check{\boldsymbol{U}}^{2;2}$ and $\check{\boldsymbol{U}}^{2;3}$ are used, we obtain agreement with the rigid sphere result (Einsteinian viscosity).

VI. Discussion

The linearity of the Stokes' equations enabled us to obtain the membrane deformation in any flow field which, at least in the vicinity of the particle, is governed by the equations of creeping motion. The analysis has been restricted, however, to small deviations from sphericity. As demonstrated, this requires (a) either a characteristic time scale for the membrane τ_0 to be small in comparison to a characteristic time scale for the flow τ (the membrane reacts almost like a rigid membrane) or (b) the inside to be highly viscous (the inside behaves almost like a solid). As far as a first order deformation is concerned, this latter case is independent of the membrane properties altogether and thus is identical to the case of a surface tension droplet.

On the other hand, (a) explicitly involves the membrane rheology. Concentrating on the case of purely elastic membranes (treated as two-dimensional limit of a three-dimensional solid), the assumption of small deformation allows us to employ the concept of linear elasticity, as first approximation. The time evolution for the tensors characterizing the in-surface and overall deformation is obtained and shown to depend on the time ratio τ_0, Poisson's ratio, and the viscocity ratio $\varkappa$. For homogeneous flows, two characteristic relaxation times are needed for the deformation tensors $\boldsymbol{U}^{2;2}$ and $\boldsymbol{U}^{2;3}$. On the other hand, for quadradic flows, three additional times enter via the time evolution of the deformation tensors $\boldsymbol{U}^{2;1}$, $\boldsymbol{U}^{3;2}$, and $\boldsymbol{U}^{3;3}$. Thus, even the first order deformation results soon become highly complex. Yet, no matter how complex the undisturbed Stokes' flow is, the steady state limits of the deformation tensors are quite simple in form (see eq. 3.39). Furthermore, to first order they are independent of the viscosity (or rheology) of the inside fluid.

This contrasts the case of surface tension droplets (surface tension σ) where, according to [20],*) the steady state shape of the droplet is given by:

$$r_s = R\left\{1 + \hat{\tau}_0 \sum_{l=2} R^{l-2}\left[H_{1l}\check{\boldsymbol{B}}^l + \frac{R^2}{\eta_0} H_{2l}\check{\boldsymbol{C}}^l\right](\cdot l)\,\hat{\boldsymbol{Y}}^l\right\} \tag{6.1}$$

with the relaxation time:

$$\hat{\tau}_0 = \frac{R\eta_0}{\sigma}. \tag{6.2}$$

The coefficients are:

$$H_{1l} = \frac{2l-1}{(l-1)(l+1)(l+2)}\,\frac{\varkappa(2l^2+4l+3)+2l(l+2)}{1+\varkappa}, \tag{6.3a}$$

$$H_{2l} = \frac{1}{(l-1)(l+1)(l+2)}\,\frac{\varkappa\left(l^2+\frac{1}{2}\right)+l^2-1}{(1+\varkappa)}. \tag{6.3b}$$

Comparing these expressions with eq. (3.30) allows us to utilize droplet results for the prediction of the first order capsule deformation (and vice versa). For example, for a droplet freely suspended in tube flow (tube radius R_0, centerline velocity v_0 in the z-direction) with b the distance of the center of the droplet shape is [20]:

$$r_s = R\left[1 + \hat{\tau}_0\frac{v_0 R}{R_0^2}\left(-\frac{19\varkappa+16}{12(1+\varkappa)}\,\frac{b}{R}\cos\varphi P_2^1(\cos\theta) + \frac{11\varkappa+10}{20(1+\varkappa)}P_3(\cos\theta)\right)\right]. \tag{6.4}$$

*) These authors admit a surface tension which depends linearly on Γ^1, the deviation concentration of surfactants from equilibrium (equilibrium value Γ_0). For Γ^1, a conservation equation of the form:

$$-K\Gamma^1 = \Gamma_0\frac{\partial}{\partial \boldsymbol{r}_s}\cdot \boldsymbol{v}_s - D_s\Delta_s{}^2\Gamma^1$$

is assumed to hold. Here, K is a (given) constant, D_s the diffusivity of surfactants on the interface, and $\boldsymbol{v}_s$ the (fluid) velocity parallel to the interface

Here, θ and φ are the polar and azimuthal angle of a spherical coordinate system centered at the particle center. The angle θ is measured relative to the z-axis while the x-axis (relative to which φ is defined) points from the tube axis to the drop center. Thus, if the drop is sufficiently far from the axis ($b/R \gg 1$), it deforms into an ellipsoid (as expected) while deviations from that shape become more pronounced the closer the particle is to the tube axis. At the axis itself ($b = 0$), it deforms into an ovoid with the pointed end in the direction of flow.

For this type of flow, only the tensors $\boldsymbol{B}^2$ and $\boldsymbol{B}^3$ matter as far as eq. (3.40) is concerned ($\boldsymbol{C}^2$ requires a flow varying cubically with position; see 3.3c). Thus, replacing H_{12} and H_{13} of (5.3a) with the values dictated by (3.40), the result for the capsule should be:

$$r_s = R\left[1 + \tau_0 \frac{v_0 R}{R_0}\left(-\frac{10}{3}(2+\nu)\frac{b}{R}\cos\varphi\, P_2^1(\cos\theta) + \frac{7}{4}(3+\nu)\, P_3(\cos\theta)\right)\right]. \tag{6.5}$$

Using:

$$\check{\boldsymbol{B}}^2 = -\frac{1}{6}\frac{v_0 b}{R_0^2}[\hat{\boldsymbol{i}}\hat{\boldsymbol{k}} + \hat{\boldsymbol{k}}\hat{\boldsymbol{i}}], \tag{6.6a}$$

$$\check{\boldsymbol{B}}^3 = \frac{1}{225}\frac{v_0}{R_0^2}\{(\hat{\boldsymbol{i}}\hat{\boldsymbol{i}} + \hat{\boldsymbol{j}}\hat{\boldsymbol{j}})\,\hat{\boldsymbol{k}} + (\hat{\boldsymbol{i}}\hat{\boldsymbol{k}}\hat{\boldsymbol{i}} + \hat{\boldsymbol{j}}\hat{\boldsymbol{k}}\hat{\boldsymbol{j}}) + \hat{\boldsymbol{k}}(\hat{\boldsymbol{i}}\hat{\boldsymbol{i}} + \hat{\boldsymbol{j}}\hat{\boldsymbol{j}}) - 2\hat{\boldsymbol{k}}\hat{\boldsymbol{k}}\hat{\boldsymbol{k}}\} \tag{6.6b}$$

and

$$\frac{1}{R}\boldsymbol{\chi} = \sin\theta\cos\varphi\,\hat{\boldsymbol{i}} + \sin\theta\sin\varphi\,\hat{\boldsymbol{j}} + \cos\theta\,\hat{\boldsymbol{k}} \tag{6.6c}$$

the validity of (5.5) is readily ascertained.

For droplets moving along the axis in tube flow, a numerical study revealed deformation into an ovoid-like shape, even for droplets almost as large as the tube [21]. Based upon the "similarity" between droplet and capsule shape as obtained in this paper, we thus expect large capsules (with respect to tube radius) to show the ovoid shape, too. This conjecture will be scrutinized in a future study.

References

1 CLIFT, R.; GRACE, J. R.; WEBER, M. E., Bubbles, Drops and Particles, Academic Press, New York (1978).
2 TAYLOR, G. I., The viscosity of a fluid containing small drops of another fluid, Proc. Roy. Soc. **A138**, 41–48 (1932).
3 COX, R. G., The deformation of a drop in a general time dependent fluid flow, J. Fluid Mech. **37**, 601–623 (1969).
4 FRANKEL, N. A.; ACRIVOS, A., The constitutive equation for a dilute emulsion, J. Fluid Mech. **44**, 67–78 (1970).
5 ROSCOE, R., On the rheology of a suspension of viscoelastic spheres in a viscous liquid, J. Fluid Mech. **28**, 273–293 (1967).
6 GODDARD, J. D.; MILLER, CH., Nonlinear effects in the rheology of dilute suspensions, J. Fluid Mech. **28**, 657–673(1967).
7 RUSHTON, E.; DAVIS, G. A., Settling of encapsulated droplets at low Reynolds numbers, Int. I. Multiphase Flow, **9**, 337–342 (1983).
8 DAVIS, A. M. J.; BRENNER, H., Emulsions containing a third solid internal phase, Proc. A.S.C.E. **107** (EM3), 609–621 (1981).
9 BRUNN, P. O., On the rheology of viscous drops surrounded by an elastic shell, Biorheol. **17**, 419–430 (1980).
10 BRUNN, P. O., The deformation of a viscous particle surrounded by an elastic shell in a general time dependent linear flow field, J. Fluid Mech. **126**, 533–544 (1983).
11 BARTHÈS-BIESEL, D., Motion of a spherical microcapsule freely suspended in a linear shear flow, J. Fluid Mech. **100**, 831–853 (1980).
12 BRENNEN, CH., A concentrated suspension model for the Couette rheology of blood, Can. J. Chem. Eng. **53**, 126–133 (1975).
13 BRUNN, P. O., The effect of Brownian motion for a suspension of spheres, Rheol. Acta **15**, 104–119 (1976).
14 BRUNN, P. O., The motion of a slightly deformed sphere in a visco-elastic fluid. Rheol. Acta **18**, 229–243 (1979).
15 BRUNN, P. O., Faxén relations of arbitrary order and their application. J. Appl. Math. Phys. (ZAMP) **31**, 332–343 (1980).
16 GUERLET, B.; BARTHÈS-BIESEL, D.; STOLTZ, J. F., Deformation of a sphered red blood cell freely suspended in a simple shear flow, INSERIM 71, 257–264 (1977).
17 HABER, S.; HETSERONI, G., The dynamics of a deformable drop suspended in an unbounded Stokes flow, J. Fluid Mech. **49**, 257–277 (1971).
18 RALLISON, J. M., Note on the Faxén relations for a particle in Stokes flow, J. Fluid Mech. **88**, 529–533 (1978).
19 BARTHÈS-BIESEL, D.; RALLISON, J. M., The time dependent deformation of a capsule freely suspended in a linear shear flow, J. Fluid Mech. **113**, 251–267 (1981).
20 HABER, S.; HETSERONI, G., Hydrodynamics of a drop submerged in an unbounded arbitrary velocity field in the presence of surfactants, Appl. Sci. Res. **25**, 215–233 (1971).
21 HYMAN, W. I.; SKALAK, R., Non-newtonian behaviour of a suspension of liquid drops in tube flow. AIChE J. 18, 149–154 (1972).

Received August 19, 1983

Address: Prof. P. O. BRUNN, Department of Chemical Engineering and Applied Chemistry, Columbia University New York, New York 10027, U.S.A.

BOOK REVIEWS

Krein, M. G., Topics in Differential and Integral-Equations and Operator Theory. Basel-Boston-Stuttgart, Birkhäuser Verlag 1983. 311 S., sFr. 62.–. ISBN 3-7643 1517-2 (Operator Theory 7) – Translation from the Russian

Einige bedeutende Arbeiten des bekannten sowjetischen Mathematikers M. G. Krein erscheinen in diesem Sammelband erstmals in englischer Sprache und werden somit einem größeren internationalen Leserkreis zugänglich gemacht. Dies ist um so mehr zu begrüßen, als es sich besonders bei den ersten beiden Beiträgen um Arbeiten von hohem mathematisch-historischen Wert handelt, die zudem noch in allgemein schwer zugänglichen Quellen veröffentlicht waren. In der ersten (aus dem Jahr 1955 stammenden) Arbeit werden Systeme linearer Differentialgleichungen der Gestalt

$$dx/dt = \lambda J H(t)\, x \tag{1}$$

betrachtet. Dabei ist x eine $2n$-dimensionale reelle Vektorfunktion, $H(t)$ eine symmetrische Matrix reeller, periodischer Funktionen und

$$J = \begin{pmatrix} 0 & I_n \\ -I_n & 0 \end{pmatrix}, \qquad I_n = \text{Einheitsmatrix}\,.$$

Für diese Gleichungssysteme, die die Schwingungsbewegungen eines mechanischen Systemes mit n Freiheitsgraden beschreiben, wird die Theorie der λ-Stabilitätszonen entwickelt. Das ist die Menge derjenigen Parameterwerte λ, für die das Gleichungssystem (1) auch bei kleinen Störungen nur beschränkte Lösungen besitzt. Diese bereits klassisch gewordene und in einschlägige Lehrbücher eingeflossene Theorie spielt bekanntlich in vielen Gebieten der Physik eine Rolle, wo die Stabilität von Schwingungsbewegungen untersucht wird.

Die zweite Arbeit (1963/64) trägt den Titel: „Über einige neuere Untersuchungen in der Störungstheorie für selbstadjungierte Operatoren". Es handelt sich hierbei um eine Arbeit mit weitreichenden Konsequenzen, denn M. G. Krein hat hier durch grundlegende Ideen, wie z. B. durch die Einführung der Spektralverschiebungsfunktion und der später nach ihm benannten Störungsdeterminante sowie durch Klärung der grundlegenden Zusammenhänge zwischen diesen Begriffen einen wesentlichen Beitrag zur Entwicklung der Quanten-Streutheorie geleistet.

Die dritte Arbeit „Über nichtlineare Integralgleichungen, die in der Theorie der Wiener-Hopf-Gleichungen eine Rolle spielen", ist jüngeren Datums (1976/77). Hier werden Wiener-Hopfsche Integralgleichungen

$$g(t) - \int_0^\infty k(t-s)\, g(s)\, ds = f(t) \qquad (0 \leq t < \infty) \tag{2}$$

mit einem Kern der Gestalt $k(t) = \int_0^\infty \exp(-|t|\, u)\, d\, z(u)$ untersucht. Solche Gleichungen besitzen einen direkten Ursprung in verschiedenen Problemen der Astrophysik, sie haben aber auch Anwendungen in ganz anderen Gebieten (z. B. in der Wahrscheinlichkeitstheorie). Durch die ausführliche Untersuchung einer nichtlinearen Integralgleichung bezüglich derjenigen Funktion, die die Faktorisierung der Fouriertransformierten des Kernes beschreibt, gelangt der Autor zu detaillierten analytischen Aussagen für die Ausgangsgleichung (2) (wie z. B. Struktur und asymptotisches Verhalten der Lösung bzw. der Lösungen), die weit über die aus der allgemeinen Theorie der Wiener-Hopf-Gleichungen folgenden Aussagen (bzgl. Lösbarkeitseigenschaften in verschiedenen Räumen in Abhängigkeit von den Eigenschaften des Maßes z) hinausgehen. Für die genannten Anwendungen ist diese Arbeit von aktuellem Interesse.

Drei kleinere Arbeiten über paarige Wiener-Hopfsche Integralgleichungen (1958), über Abschätzungen für die charakteristischen Zahlen linearer Integralgleichungen mit glatten Kernen (1970) und über die S-Matrizen kanonischer Differentialgleichungssysteme (1968), die M. G. Krein zusammen mit seinen Schülern I. C. Gohberg bzw. F. E. Melik-Adamyan verfaßt hat, ergänzen das Sammelwerk.

Alle Arbeiten sind – und das ist besonders hervorzuheben – sehr verständlich und elegant in der Darstellungsweise. Sie sind self-contained und mit Grundkenntnissen der linearen Algebra, der Funktionalanalysis und der Funktionentheorie erschließbar. Druckfehler waren nur wenige zu finden, sie sind unerheblich.

Alles in allem handelt es sich um ein Buch, dessen Lektüre dem Leser Vergnügen bereiten wird sowie manche Inspiration vermitteln kann und Hochachtung vor der Vielseitigkeit und dem Gedankenreichtum M. G. Kreins abnötigt.

Berlin A. Pomp

Colton, D. / Kress, R., Integral Equation Methods in Scattering Theory. New York et al., John Wiley & Sons 1983. XII, 271 S., £ 35.75. ISBN 0 471 86420-X

In dem Buch werden Randwertprobleme (Dirichletsche und Neumannsche) für die (skalare und vektorielle) Helmholtz-Gleichung sowie für die zeitharmonischen Maxwell-Gleichungen behandelt und zwar mit Hilfe von Integralgleichungsmethoden. Zur Lösung der äußeren (Dirichletschen und Neumannschen) Probleme werden iterative Methoden herangezogen. Das Verhalten der Lösungen für niedrige Frequenzen wird untersucht.

Schließlich wird das Problem der Bestimmung von Hindernissen (für die Wellenausbreitung) aus dem asymptotischen Verhalten der Lösungen aufgeworfen und zwar für exakte und für näherungsweise gegebene Daten. Dieses Problem hängt mit dem sog. inversen Streuproblem zusammen (allerdings ist in diesem Buch nur am Rande von Streutheorie die Rede; Streuoperator und Streumatrix werden nicht eingeführt, lediglich der Streuquerschnitt wird definiert).

Das Buch enthält einleitend Abschnitte über die Rieß-Fredholmsche Theorie kompakter Operatoren und über Regularitätseigenschaften von Flächenpotentialen sowie im Text einen Exkurs über ganze Funktionen vom Exponentialtyp.

Berlin H. Baumgärtel

Aubin, J. P. / Bensoussan, A. / Ekeland, I. (eds.), Advances in Hamiltonian Systems. Boston-Basel-Stuttgart, Birkhäuser Verlag 1983. VIII, 195 S., sFr. 46.–. ISBN 3-7643-3130-5 (Annals of the Ceremade 2)

Der vorliegende zweite Band der „Annals of the CEREMADE" enthält die Ergebnisse einer Konferenz über Hamiltonsche Systeme, die 1981 in Rom stattgefunden hat. Er dokumentiert eindrucksvoll, welche Fortschritte in der letzten Dekade bei der Untersuchung periodischer Lösungen der Hamiltonschen Gleichungen (und damit von nichtlinearen Wellengleichungen) erzielt wurden. In den klassischen Arbeiten auf diesem Gebiet (Poincaré, Ljapunov, Kolmogorov, Arnold, Moser) beschränkte man sich auf Aussagen lokaler Natur, d. h. es wurden Lösungen in der Nähe von Gleichgewichtslagen betrachtet. Das Hauptinteresse der Autoren des vorliegenden Bandes richtet sich dagegen auf das Auffinden globaler Lösungen. Es werden Existenzsätze angegeben und Fragen nach der Anzahl der Lösungen zu einem gegebenen Energieniveau oder nach der Existenz einer minimalen Periode gestellt.

Die charakteristische Schwierigkeit dieser Untersuchungen liegt in der Tatsache begründet, daß die Variationsformulierung für die Hamiltonschen Gleichungen auf ein nichtkonvexes Funktional führt, welches weder nach unten noch nach oben beschränkt ist. Hier läßt sich die Existenz kritischer Punkte nicht mehr aus dem Vorhandensein eines globalen Minimums bzw. Maximums ableiten. Durch das Aufstellen dualer Variationsprinzipien für solche unbeschränkten nichtkonvexen Funktionale gelingt es dennoch, zu den gewünschten Resultaten zu gelangen. Hierbei werden auf äußerst fruchtbare Weise topologische Methoden und Ergebnisse der nichtlinearen Funktionalanalysis bzw. der konvexen Analysis kombiniert.

Im Einzelnen enthält der Band folgende Beiträge:

A. Ambrosetti: Recent advances in the study of the existence of periodic orbits of Hamiltonian systems; V. Benci: The direct method in the study of periodic solutions of Hamiltonian systems with prescribed period; G. Mancini: Periodic solutions of Hamiltonian systems having prescribed minimal period; I. Ekeland and J.-M. Lasry: Duality in non convex variational problems; P. Bernhard: La théorie de la seconde variation et le problème linéaire quadratique; J. Blot: Méthodes asymptotiques dans l'étude de systèmes hamiltoniens non autonomes; E. Gaussens: Numerical research of periodic solutions for a Hamiltonian system.

Berlin E. Krauss

ZAMM · Z. Angew. Math. u. Mech. 65 (1985) 1, 13–17

OLEINIK, O. A.; SHAMAEV, A. S.; YOSIFIAN, G. A.

On the Convergence of the Energy, Stress Tensors, and Eigenvalues in Homogenization Problems of Elasticity

Dedicated to the 60th birthday of K. NICKEL

In dem vorliegenden Artikel wird die Konvergenz von Energieintegralen, Spannungstensoren und den Frequenzen der freien Schwingungen für inhomogene und poröse elastische Körper mit einer periodischen Struktur der Periode ε bei $\varepsilon \to 0$ untersucht. Die Ergebnisse beruhen auf Abschätzungen der Lösungen für Randwertprobleme eines elastischen Systems mit schnell oszillierenden periodischen Koeffizienten, wie sie in [4] *und* [11] *erhalten wurden.*

In this paper we study the convergence of energy integrals, stress tensors and frequencies of free vibrations for non-homogeneous and porous elastic bodies with a periodic structure of period ε as $\varepsilon \to 0$. The results are based on the estimates for solutions of the boundary value problem of the elasticity system with rapidly oscillating periodic coefficients, obtained in [4], [11].

В этой статье изучается сходимость интегралов энергии, тензоров напряжений и частот свободных колебаний для неоднородных и пористых упругих тел периодической структуры с периодом ε при $\varepsilon \to 0$. Результат основывается на оценке решений краевой задачи для системы теории упругости с быстро осциллирующими периодическими коэффициентами, полученной в [4], [11].

Many problems in physics and mechanics lead to the study of processes in non-homogeneous and porous media with a periodic structure. As an example one can consider an elastic body consisting of several materials with different properties, or an elastic body with periodically situated cavities (holes). Differential equations which describe homogenized macroscopic properties of such non-homogeneous and porous media have been obtained in a number of papers, together with the estimates, characterizing the difference of the solutions of these homogenized equations and the solutions of the equations having rapidly oscillating coefficients and describing composite materials with cavities, (see [1]–[11]).

It is necessary in some problems of mechanics to find such quantities as energy, stress tensors and frequencies of free vibrations of non-homogeneous elastic bodies with a periodic structure. One can approximate these quantities using the related homogenized boundary value problem, and it is important to estimate the error.

The problem of the convergence of the energy integrals for second order elliptic equations with periodic rapidly oscillating coefficients was first considered in [1]. The convergence of the energy integrals for strongly G-convergent sequences of elliptic equations of any order was proved in [2]. The system of elasticity with rapidly oscillating periodic coefficients in domains with cavities is considered in papers [4]–[8]. In [2], [9], [10] one can find a large number of references related to the G-convergence, homogenization problems and asymptotic analysis of periodic structures.

In this paper we study the convergence of energy integrals, stress tensors and frequencies of free vibrations for non-homogeneous and porous elastic bodies with a periodic structure of period ε as $\varepsilon \to 0$. The results are based on the estimates for solutions of the boundary value problem of the elasticity system with rapidly oscillating periodic coefficients, obtained in [4], [11].

Consider a set G^0 belonging to the unit cube $Q = \{\xi \in \mathbb{R}^n: 0 < \xi_j < 1, j = 1, \dots, n\}$ and consisting of a finite number of mutually non-intersecting smooth domains. Suppose that the distance between G^0 and ∂Q is positive. Denote by $X + z$ the set $\{y \in \mathbb{R}^n: y = x + z, x \in X\}$ and by εX the set $\{x \in \mathbb{R}^n: \varepsilon^{-1}x \in X\}$. Let $G_1 = \bigcup_{z \in \mathbb{Z}^n} (G^0 + z)$ where $\mathbb{Z}^n$ is the set of vectors $z = (z_1, \dots, z_n)$ with integer components, $G_\varepsilon = \varepsilon G_1$, $\Omega^\varepsilon = \Omega \setminus \bar{G}_\varepsilon$, $S_\varepsilon = \partial\Omega^\varepsilon \setminus \partial\Omega$, $\Gamma_\varepsilon = \partial\Omega \cap \partial\Omega^\varepsilon$, Ω — is a domain with a smooth boundary. The set Ω^ε belongs to a class of perforated domains.

Let $u = (u_1, \dots, u_n)^*$ be a vector-column with components $u_1, \dots, u_n$. Denote by $H^1(\omega)$ the space of vector valued functions with the norm

$$\|u\|_{H^1(\omega)} = \left(\int_\omega \left(\sum_{j=1}^n u_j^2 + \sum_{i,j=1}^n \left|\frac{\partial u_j}{\partial x_i}\right|^2\right) dx\right)^{1/2}.$$

Consider the following boundary value problem of elasticity

$$\left.\begin{aligned} L_\varepsilon(u^\varepsilon) &\equiv \frac{\partial}{\partial x_h}\left(C^{hk}\left(\frac{x}{\varepsilon}\right)\frac{\partial u^\varepsilon}{\partial x_k}\right) = f(x) \quad \text{in } \Omega^\varepsilon, \\ u^\varepsilon = 0 \quad \text{on} \quad &\Gamma_\varepsilon, \sigma_\varepsilon(u^\varepsilon) \equiv C^{hk}\left(\frac{x}{\varepsilon}\right)\frac{\partial u^\varepsilon}{\partial x_k}\nu_h = 0 \quad \text{on} \quad S_\varepsilon, \end{aligned}\right\} \tag{1}$$

where $C^{hk}(\xi)$ are $(n \times n)$ matrices whose elements $c_{ij}^{hk}(\xi)$ belong to $L^\infty(\mathbb{R}^n_\xi)$ and are periodic functions in $\xi_1, \dots, \xi_n$ of period 1 (1-periodic in ξ), $u^\varepsilon = (u_1^\varepsilon, \dots, u_n^\varepsilon)^*$, $f = (f_1, \dots, f_n)^*$, $f \in C^{1+\alpha}(\bar{\Omega})$, $0 < \alpha < 1$; $\nu = (\nu_1, \dots, \nu_n)$ is the exterior normal vector to S_ε. Throughout the paper summation over repeated indices from 1 to n is assumed. As usually we assume that the coefficients $c_{ij}^{hk}(\xi)$ satisfy the following conditions

$$c_{ij}^{hk}(\xi) = c_{ji}^{kh}(\xi) = c_{hj}^{ik}(\xi), \qquad \xi \in Q \setminus G^0, \tag{2}$$

and for any $\xi \in Q \setminus G^0$ and any symmetric $(n \times n)$ matrice $\eta = \{\eta_s^t\}$

$$\varkappa_1 \eta_i^h \eta_i^h \leqq c_{ij}^{hk}(\xi)\, \eta_i^h \eta_j^k \leqq \varkappa_2 \eta_i^h \eta_i^h\,, \tag{3}$$

where $\varkappa_1, \varkappa_2$ are positive constants.

Set $(u, v) = u_i v_i$.

We say that u^ε is a weak solution of problem (1) if $u^\varepsilon \in H^1(\Omega^\varepsilon)$, $u^\varepsilon|_{\Gamma_\varepsilon} = 0$ and if the integral identity

$$-\int\limits_{\Omega^\varepsilon} \left(C^{hk}\left(\frac{x}{\varepsilon}\right) \frac{\partial u^\varepsilon}{\partial x_k}, \frac{\partial v}{\partial x_h}\right) dx = \int\limits_{\Omega^\varepsilon} (f, v)\, dx$$

holds for any $v \in H^1(\Omega^\varepsilon)$, $v|_{\Gamma_\varepsilon} = 0$.

We denote by $\hat{H}^1(Q \setminus G^0)$ the completion of the space of 1-periodic in ξ and infinitely differentiable in $\mathbb{R}^n \setminus G_1$ vector valued functions, with respect to the norm of $H^1(Q \setminus G^0)$. A matrix M is said to belong to $\hat{H}^1(Q \setminus G^0)$ if its columns belong to this space.

Let the matrices $N_p(\xi)$ $(p = 1, \dots, n)$ be 1-periodic in ξ weak solutions of the following boundary value problems

$$\left.\begin{aligned} &\frac{\partial}{\partial \xi_k}\left(C^{kj}(\xi) \frac{\partial}{\partial \xi_j} N_p(\xi)\right) = -\frac{\partial}{\partial \xi_k} C^{kp}(\xi) \quad \text{in} \quad \mathbb{R}^n \setminus G_1\,, \\ &\sigma(N_p) = -C^{kp}\nu_k \quad \text{on} \quad \partial G_1, \int\limits_{Q\setminus G^0} N_p(\xi)\, d\xi = 0\,. \end{aligned}\right\} \tag{4}$$

A matrix $N_p(\xi)$ is called a weak solution of problem (4) if $N_p(\xi) \in \hat{H}^1(Q \setminus G^0)$, $\int\limits_{Q\setminus G^0} N_p(\xi)\, d\xi = 0$ and the integral identity

$$-\int\limits_{Q\setminus G^0} C^{kj}(\xi) \frac{\partial N_p(\xi)}{\partial \xi_j} \frac{\partial M}{\partial \xi_k} d\xi = \int\limits_{Q\setminus G^0} C^{kp}(\xi) \frac{\partial M}{\partial \xi_k} d\xi$$

holds for any matrix $M \in \hat{H}^1(Q \setminus G^0)$.

Set

$$h^{pq} = \left(\operatorname{mes}(Q \setminus G^0)\right)^{-1} \int\limits_{Q\setminus G^0} \left(C^{pq}(\xi) + C^{pj}(\xi) \frac{\partial N_q}{\partial \xi_j}\right) d\xi\,, \qquad p, q = 1, \dots, n$$

and consider the following boundary value problem

$$\hat{L}(U) \equiv \frac{\partial}{\partial x_p}\left(h^{pq} \frac{\partial U}{\partial x_q}\right) = f(x) \quad \text{in} \quad \Omega\,, \qquad U = 0 \quad \text{on} \quad \partial\Omega\,. \tag{5}$$

It is known (see [8]) that system (5) is of the elasticity type, i.e. the components h_{ij}^{pq} of matrices h^{pq} satisfy conditions similar to (2), (3).

The following theorem gives estimates for the difference between the solutions of problems (1) and (5). The proof can be found in [11] (see also [4]).

THEOREM 1: *For solutions u^ε and U of problems (1) and (5) the following estimates hold*

$$\|u^\varepsilon - U\|_{L^2(\Omega^\varepsilon)} \leqq c\sqrt{\varepsilon}\,, \tag{6}$$

$$\left\|u^\varepsilon - U - \varepsilon N_p\left(\frac{x}{\varepsilon}\right) \frac{\partial U}{\partial x_p}\right\|_{H^1(\Omega^\varepsilon)} \leqq c\sqrt{\varepsilon}\,, \tag{7}$$

where c is a constant independent of ε.

These estimates allow to study the convergence of energy integrals, stress tensors and frequencies of free vibrations related to (1) as $\varepsilon \to 0$.

THEOREM 2 (On the convergence of energy integrals): *Suppose that Ω' is a subdomain of Ω with the boundary in C^1, $f \in C^{1+\gamma}(\bar{\Omega})$, $0 < \gamma < 1$. Then*

$$\left|\int\limits_{\Omega^\varepsilon \cap \Omega'} \left(C^{jk}\left(\frac{x}{\varepsilon}\right) \frac{\partial u^\varepsilon}{\partial x_k}, \frac{\partial u^\varepsilon}{\partial x_j}\right) dx - \operatorname{mes}(Q \setminus G^0) \int\limits_{\Omega'} \left(h^{jk} \frac{\partial U}{\partial x_k}, \frac{\partial U}{\partial x_j}\right) dx\right| \leqq c\sqrt{\varepsilon}\,, \tag{8}$$

where c = const and does not depend on ε.

Proof. It follows from theorem 1 that $u^\varepsilon(x) = U(x) + \varepsilon N_p(x/\varepsilon)\, \partial U/\partial x_p + r_\varepsilon(x)$, where $\|r_\varepsilon(x)\|_{H^1(\Omega^\varepsilon)} \leqq c_1\sqrt{\varepsilon}$. Hence

$$\frac{\partial u^\varepsilon}{\partial x_i} = \frac{\partial U}{\partial x_i} + \frac{\partial N_p(x/\varepsilon)}{\partial \xi_i} \frac{\partial U}{\partial x_p} + q_i^\varepsilon(x), \ \|q_i^\varepsilon(x)\|_{L^2(\Omega^\varepsilon)} \leqq c_2\sqrt{\varepsilon}\,, \qquad c_2 = \text{const.} \tag{9}$$

Therefore

$$\int\limits_{\Omega^\varepsilon \cap \Omega'} \left(C^{ij}\left(\frac{x}{\varepsilon}\right) \frac{\partial u^\varepsilon}{\partial x_j}, \frac{\partial u^\varepsilon}{\partial x_i}\right) dx = \int\limits_{\Omega^\varepsilon \cap \Omega'} \left(C^{ij}\left(\frac{\partial U}{\partial x_j} + \frac{\partial N_t}{\partial \xi_j} \frac{\partial U}{\partial x_t}\right), \frac{\partial U}{\partial x_i} + \frac{\partial N_s}{\partial \xi_i} \frac{\partial U}{\partial x_s}\right) dx +$$

$$+ p^\varepsilon = \int\limits_{\Omega^\varepsilon \cap \Omega'} \left(\frac{\partial}{\partial \xi_i} (N_s^* + \xi_s E)\, C^{ij} \frac{\partial}{\partial \xi_j} (N_t + \xi_t E) \frac{\partial U}{\partial x_t}, \frac{\partial U}{\partial x_s}\right) dx + p^\varepsilon\,, \tag{10}$$

where $|p^\varepsilon| \leqq c\sqrt{\varepsilon}$, $\xi = \varepsilon^{-1}x$, E is the unit matrix, N_t^* is the conjugate matrix to N_t. It is shown in [8] that

$$h^{st} = (\text{mes}\,(Q \setminus G^0))^{-1} \int\limits_{Q\setminus G^0} \frac{\partial}{\partial \xi_i} (N_s^* + \xi_s E)\, C^{ij}(\xi)\, \frac{\partial}{\partial \xi_j} (N_t + \xi_t E)\, d\xi\,. \tag{11}$$

Denote by B^ε the set in $\mathbb{R}^n$ consisting of all cubes $n_\alpha^\varepsilon(\alpha = 1, \dots, N(\varepsilon))$ of the form $\varepsilon(Q + z)$, such that $n_\alpha^\varepsilon \subset \Omega'$. Denote by $\mathcal{D}^\varepsilon$ the set in $\mathbb{R}^n$ consisting of all cubes $m_\beta^\varepsilon(\beta = 1, \dots, M(\varepsilon))$ of the form $\varepsilon(Q + z)$ and such that $m_\beta^\varepsilon \cap \partial\Omega' \neq \emptyset$. Let x_ε^α be the center of n_α^ε. Set

$$M^{st}(\xi) \equiv \frac{\partial}{\partial \xi_i} (N_s^*(\xi) + \xi_s E)\, C^{ij}(\xi) \frac{\partial}{\partial \xi_j} (N_t(\xi) + \xi_t E)\,,$$

$$\widetilde{M}^{st}(\xi) = 0 \quad \text{in} \quad \varepsilon G_1\,, \qquad \widetilde{M}^{st}(\xi) = M^{st}(\xi) \quad \text{in} \quad \mathbb{R}^n \setminus \varepsilon G_1\,,$$

$$H^{st}(\xi) = \widetilde{M}^{st}(\xi) - \text{mes}\,(Q \setminus G^0)\, h^{st}\,, \qquad \Psi_t(x) \equiv \frac{\partial U(x)}{\partial x_t}.$$

It follows from (10) that

$$\int\limits_{\Omega^\varepsilon \cap \Omega'} \left(C^{ij}\left(\frac{x}{\varepsilon}\right) \frac{\partial u^\varepsilon}{\partial x_j}, \frac{\partial u^\varepsilon}{\partial x_i}\right) dx - \text{mes}\,(Q \setminus G^0) \int\limits_{\Omega'} \left(h^{st} \frac{\partial U(x)}{\partial x_t}, \frac{\partial U(x)}{\partial x_s}\right) dx = \int\limits_{\Omega'} \left(H^{st}\left(\frac{x}{\varepsilon}\right) \Psi_t(x), \Psi_s(x)\right) dx + p^\varepsilon =$$

$$= \sum_{\alpha=1}^{N(\varepsilon)} \int\limits_{n_\alpha^\varepsilon} \left(H^{st}\left(\frac{x}{\varepsilon}\right) \Psi_t(x_\varepsilon^\alpha), \Psi_s(x_\varepsilon^\alpha)\right) dx + \sum_{\alpha=1}^{N(\varepsilon)} \int\limits_{n_\alpha^\varepsilon} \left[\left(H^{st}\left(\frac{x}{\varepsilon}\right) \Psi_t(x), \Psi_s(x)\right) - \left(H^{st}\left(\frac{x}{\varepsilon}\right) \Psi_t(x_\varepsilon^\alpha), \Psi_s(x_\varepsilon^\alpha)\right)\right] dx +$$

$$+ \int\limits_{\mathcal{D}^\varepsilon \cap \Omega'} \left(H^{st}\left(\frac{x}{\varepsilon}\right) \Psi_t(x), \Psi_s(x)\right) dx + p^\varepsilon \equiv I_1 + I_2 + I_3 + p^\varepsilon\,.$$

It is easy to see that $I_1 = 0$ due to (11). Let us estimate I_2, I_3. Since $\partial\Omega$ and $\partial\Omega'$ are smooth we have $M(\varepsilon) \leqq c_1 \varepsilon^{1-n}$, $N(\varepsilon) \leqq c_2 \varepsilon^{-n}$, and $|\Psi_s(x_\varepsilon^\alpha) - \Psi_s(x)| \leqq c_3 \varepsilon$ for $x \in n_\alpha^\varepsilon$, because $f \in C^{1+\gamma}(\bar{\Omega})$ and therefore $U \in C^{3+\gamma}(\bar{\Omega})$. We thus obtain that

$$|I_2| = \left|\sum_{\alpha=1}^{N(\varepsilon)} \int\limits_{n_\alpha^\varepsilon} \left(H^{st}\left(\frac{x}{\varepsilon}\right) (\Psi_t(x) - \Psi_t(x_\varepsilon^\alpha)), \Psi_s(x)\right) dx + \sum_{\alpha=1}^{N(\varepsilon)} \int\limits_{n_\alpha^\varepsilon} \left(H^{st}\left(\frac{x}{\varepsilon}\right) \Psi_t(x_\varepsilon^\alpha), \Psi_s(x) - \Psi_s(x_\varepsilon^\alpha)\right) dx\right| \leqq$$

$$\leqq c_4 \varepsilon^{-n} \cdot \varepsilon \cdot \varepsilon^n\,,$$

$$|I_3| \leqq \sum_{\beta=1}^{M(\varepsilon)} \int\limits_{m_\beta^\varepsilon \cap \Omega'} \left|\left(H^{st}\left(\frac{x}{\varepsilon}\right) \Psi_t(x), \Psi_s(x)\right)\right| dx \leqq c_5 \varepsilon^{1-n} \cdot \varepsilon^n\,.$$

The theorem is proved.

Set

$$\sigma_\varepsilon^p(x) \equiv C^{pk}\left(\frac{x}{\varepsilon}\right) \frac{\partial u^\varepsilon}{\partial x_k}\,, \qquad \sigma^p(x) \equiv h^{pq} \frac{\partial U}{\partial x_q}.$$

The components of $\sigma_\varepsilon^p(x)$ and $\sigma^p(x)$ are equal to the components of the stress tensors, corresponding to problems (1) and (5). We define matrices $a^{ij}(\xi)$ by the formula

$$a^{ij}(\xi) \equiv C^{ij}(\xi) + C^{ik}(\xi) \frac{\partial N_j(\xi)}{\partial \xi_k} - h^{ij}\,.$$

Denote by $\tilde{\sigma}_\varepsilon^p$ the vector valued function such that

$$\tilde{\sigma}_\varepsilon^p = \sigma_\varepsilon^p \quad \text{in} \quad \Omega^\varepsilon, \quad \text{and} \quad \tilde{\sigma}_\varepsilon^p = 0 \quad \text{in} \quad \Omega \setminus \Omega^\varepsilon\,.$$

THEOREM 3 (On the convergence of stress tensors): *The vectors $\sigma_\varepsilon^p(x)$ and $\sigma^p(x)$ satisfy the inequality*

$$\left\|\sigma_\varepsilon^p(x) - \sigma^p(x) - a^{pj}\left(\frac{x}{\varepsilon}\right) \frac{\partial U}{\partial x_j}\right\|_{L^2(\Omega^\varepsilon)} \leqq c\sqrt{\varepsilon}\,, \tag{12}$$

where $c = const$ and does not depend on ε. Moreover $\tilde{\sigma}_\varepsilon^p(x) \to \text{mes}\,(Q \setminus G^0)\, \sigma^p(x)$ weakly in $L^2(\Omega)$ as $\varepsilon \to 0$.

Proof. Taking into account (7) and (9) we obtain

$$\sigma_\varepsilon^p(x) - \sigma^p(x) - a^{pj}\left(\frac{x}{\varepsilon}\right) \frac{\partial U}{\partial x_j} =$$

$$= C^{pk}\left(\frac{x}{\varepsilon}\right)\left[\frac{\partial U}{\partial x_k} + \frac{\partial N_s}{\partial \xi_k} \frac{\partial U}{\partial x_s}\right] - h^{pq} \frac{\partial U}{\partial x_q} - \left[C^{pj}(\xi) + C^{pk}(\xi) \frac{\partial N_j}{\partial \xi_k} - h^{pj}\right] \frac{\partial U}{\partial x_j} + C^{pk}\left(\frac{x}{\varepsilon}\right) q_k^\varepsilon(x) =$$

$$= C^{pk}\left(\frac{x}{\varepsilon}\right) q_k^\varepsilon(x)\,.$$

Since $|c_{ij}^{pk}(\xi)| \leqq K_1 = \text{const}$, it follows that

$$\left\| C^{pk}\left(\frac{x}{\varepsilon}\right) q_k^{\varepsilon}(x) \right\|_{L^2(\Omega^\varepsilon)} \leqq K_2 \sqrt{\varepsilon} ,$$

where K_1 and K_2 are constants independent of ε. We also have

$$\tilde{\sigma}_\varepsilon^p - \text{mes}\,(Q \setminus G^0)\, \sigma^p = A^{pq}\left(\frac{x}{\varepsilon}\right) \frac{\partial U(x)}{\partial x_q} + r_\varepsilon^p(x) ,$$

where

$$\|\overset{\circ}{r}{}_\varepsilon^p\|_{L^2(\Omega^\varepsilon)} \leqq c_1 \sqrt{\varepsilon},\ A^{pq}(\xi) = C^{pq}(\xi) + C^{ps}(\xi) \frac{\partial N_q}{\partial \xi_s} - \text{mes}\,(Q \setminus G^0)\, h^{pq}$$

for

$$\xi \in Q \setminus G^0, \qquad A^{pq}(\xi) = 0 \quad \text{for} \quad \xi \in G^0 .$$

Due to the definition of h^{pq} the matrix $A^{pq}(\xi)$ satisfies the condition $\int_Q A^{pq}(\xi)\, d\xi = 0$ and therefore $A^{pq}(\xi) \times \partial U(x)/\partial x_q \to 0$ weakly in $L^2(\Omega)$ as $\varepsilon \to 0$. The theorem is proved.

Now we consider the following eigenvalue problem

$$L_\varepsilon(u^{\varepsilon,k}) = \lambda_k^\varepsilon \varrho\left(\frac{x}{\varepsilon}\right) u^{\varepsilon,k} \ \text{in} \ \ \Omega^\varepsilon,\ u^{\varepsilon,k}|_{\Gamma_\varepsilon} = 0\ ,\ \sigma_\varepsilon(u^{\varepsilon,k})|_{S_\varepsilon} = 0\ , \tag{13}$$

$$\|u^{\varepsilon,k}\|_{L^2(\Omega^\varepsilon)} = 1\ , \qquad 0 > \lambda_1^\varepsilon \geqq \dots \geqq \lambda_k^\varepsilon \geqq \dots .$$

In the sequence $\{\lambda_k^\varepsilon\}$ each eigenvalue is counted as many times as its multiplicity, $\varrho(\xi)$ is a 1-periodic in ξ measurable bounded function such that $\varrho(\xi) \geqq \varrho_0 = \text{const} > 0$.

We also consider the eigenvalue problem for the homogenized operator

$$\hat{L}(u^k) = \lambda_k \hat{\varrho} u^k \quad \text{in} \quad \Omega\ , \qquad u^k|_{\partial\Omega} = 0\ , \tag{14}$$

$$\|u^k\|_{L^2(\Omega)} = 1\ , \qquad 0 > \lambda_1 \geqq \lambda_2 \geqq \dots\ , \qquad \hat{\varrho} = [\text{mes}\,(Q \setminus G^0)]^{-1} \int_{Q \setminus G^0} \varrho(\xi)\, d\xi\ .$$

The estimate of the difference between eigenvalues of problems (13), (14) is given by

THEOREM 4: *Suppose that one of the following conditions is satisfied*

1) $G^0 = \emptyset$ *(absence of cavities)*,
2) $\varrho(\xi) \equiv 1$ *(the density is constant)*.

Then

$$|\lambda_k^\varepsilon - \lambda_k| \leqq c(k) \sqrt{\varepsilon}\ , \tag{15}$$

where $c(k)$ is a positive constant independent of ε.

The proof of this theorem is based on theorem 1 and on some results concerning the convergence of operators in a Hilbert space (see [12]). In the case $\varrho \equiv 1$ theorem 4 is proved in papers [4], [5]. Here we give the proof for the case $G^0 = \emptyset$. We then have $\Omega = \Omega^\varepsilon$. Let $A_\varepsilon, \hat{A}: L^2(\Omega) \to L^2(\Omega)$ be operators mapping a function $f \in L^2(\Omega)$ into u^ε, U respectively, where u^ε is the solution of problem (1) and U is the solution of problem (5). According to theorem 1 for any $f \in C^{1+\alpha}(\bar{\Omega})$ we have

$$\|(A_\varepsilon - \hat{A}) f\|_{L^2(\Omega)} \leqq c \sqrt{\varepsilon}\ . \tag{16}$$

Denote by R_ε and $\hat{R}$ operators in $L^2(\Omega)$ defined by the formulas

$$R_\varepsilon u \equiv \varrho\left(\frac{x}{\varepsilon}\right) u\ , \qquad \hat{R} u \equiv \hat{\varrho} u\ , \qquad u \in L^2(\Omega)\ .$$

It is easy to see that $\lambda_k^\varepsilon = [\mu_k(\varepsilon)]^{-1}$, $\lambda_k = [\mu_k]^{-1}$, where $\mu_k(\varepsilon)$ and μ_k are eigenvalues of operators $A_\varepsilon R_\varepsilon$ and $\hat{A}\hat{R}$ respectively. In order to estimate $|\mu_k(\varepsilon) - \mu_k|$ we use the results of papers [12], [4].

Suppose that $B_m: H \to H$ is a sequence of operators in a Hilbert space H, satisfying the conditions

a) $B_m x \to B x$ in H as $m \to \infty$ for any $x \in H$,
$B_m, B: H \to H$ are compact operators;

b) images $B_m x$ of the unit sphere $\|x\|_H = 1$ belong to the same compact set $K \subset H$ for all m.

Under these conditions it is proved (see [12], [4]) that if $\mu \neq 0$ is an eigenvalue of B, then there exists a sequence $\{\mu_m\}$ (μ_m is an eigenvalue of B_m), which converges to μ as $m \to \infty$. On the other hand if a sequence μ_m (μ_m being an eigenvalue of B_m) has a limit point $\mu \neq 0$, it follows that μ is an eigenvalue of operator B. In this case one also has the inequality

$$|\mu_m - \mu| \leqq c \left| \left((B_m - B)\, v^m, v \right) \right|\ , \tag{17}$$

where $c = \text{const}$ does not depend on m; v^m and v are some eigenfunctions of operators B_m and B, corresponding to the eigenvalues μ_m and μ respectively, $\|v^m\|_H = \|v\|_H = 1$.

Operators $\{A_\varepsilon \cdot R_\varepsilon\}$ as $\varepsilon \to 0$ satisfy conditions a), b). Indeed, for any $u \in L^2(\Omega)$ we have

$$(A_\varepsilon \cdot R_\varepsilon - \hat{A}\hat{R})\, u = (A_\varepsilon - \hat{A})\, R_\varepsilon u + \hat{A}(R_\varepsilon - \hat{R})\, u\,.$$

It is easy to see that $(R_\varepsilon - \hat{R})\, u \to 0$ weakly in $L^2(\Omega)$ as $\varepsilon \to 0$. Due to the compactness of operator $\hat{A}$ we have $\hat{A}(R_\varepsilon - \hat{R})\, u \to 0$ as $\varepsilon \to 0$ in the norm of $L^2(\Omega)$. It is shown in [10] that

$$\|A_\varepsilon - \hat{A}\|_{\mathcal{L}(L^2(\Omega),\, L^2(\Omega))} \to 0 \quad \text{as} \quad \varepsilon \to 0\,. \tag{18}$$

Therefore

$$\|(A_\varepsilon - \hat{A})\, R_\varepsilon u\|_{L^2(\Omega)} \leqq \|A_\varepsilon - \hat{A}\|_{\mathcal{L}(L^2(\Omega),\, L^2(\Omega))}\, \|R_\varepsilon u\|_{L^2(\Omega)} \to 0 \quad \text{as} \quad \varepsilon \to 0\,.$$

We have thus established that condition a) is satisfied. Condition b) is a direct consequence of the fact that $\|u^\varepsilon\|_{H^1(\Omega)} \leqq C\|f\|_{L^2(\Omega)}$, where C is a constant independent of ε.

It follows from [12] (see also [4]) that

$$\mu_k(\varepsilon) \to \mu_k \quad \text{as} \quad \varepsilon \to 0 \qquad (k = 1, 2, \ldots)\,.$$

By virtue of (17) we obtain

$$\begin{aligned} |\mu_k(\varepsilon) - \mu_k| &\leqq C|(A_\varepsilon R_\varepsilon - \hat{A}\hat{R})\, u^{\varepsilon,k}, u^k)| \leqq \\ &\leqq C|((A_\varepsilon R_\varepsilon - \hat{A}R_\varepsilon)\, u^{\varepsilon,k}, u^k)| + C|((\hat{A}R_\varepsilon - \hat{A}\hat{R})\, u^{\varepsilon,k}, u^k)| = \\ &= C|(R_\varepsilon u^{\varepsilon,k}, (A_\varepsilon - \hat{A})\, u^k)| + C|((R_\varepsilon - \hat{R})\, u^{\varepsilon,k}, \hat{A}u^k)|\,, \end{aligned} \tag{19}$$

since A_ε and $\hat{A}$ are self-adjoint operators in $L^2(\Omega)$. Using (16) and the Cauchy inequality we get

$$|(R_\varepsilon u^{\varepsilon,k}, (A_\varepsilon - \hat{A})\, u^k)| \leqq \|R_\varepsilon u^{\varepsilon,k}\|_{L^2(\Omega)}\, \|(A_\varepsilon - \hat{A})\, u^k\|_{L^2(\Omega)} \leqq c_1\sqrt{\varepsilon}\,, \tag{20}$$

where constant c_1 does not depend on ε. We also have

$$\begin{aligned} |((R_\varepsilon - \hat{R})\, u^{\varepsilon,k}, \hat{A}u^k)| &= \left|\int_\Omega \left(\left(\varrho\left(\frac{x}{\varepsilon}\right) - \hat{\varrho}\right) u^{\varepsilon,k}(x)\,,\, \hat{A}u^k(x)\right) dx\right| = \left|\varepsilon \int_\Omega \left(\left(\frac{\partial}{\partial x_i} g^i\left(\frac{x}{\varepsilon}\right)\right) u^{\varepsilon,k}(x), \hat{A}u^k(x)\right) dx\right| \leqq \\ &\leqq \varepsilon \left|\int_\Omega \left(g^i\left(\frac{x}{\varepsilon}\right) \frac{\partial u^{\varepsilon,k}(x)}{\partial x_i}\,,\, \hat{A}u^k(x)\right) dx\right| + \varepsilon \left|\int_\Omega \left(g^i\left(\frac{x}{\varepsilon}\right) u^{\varepsilon,k}(x), \frac{\partial \hat{A}u^k(x)}{\partial x_i}\right) dx\right| \leqq \\ &\leqq \varepsilon c_2\big(\|u^{\varepsilon,k}(x)\|_{H^1(\Omega)}\, \|\hat{A}u^k(x)\|_{L^2(\Omega)} + \|u^{\varepsilon,k}(x)\|_{L^2(\Omega)}\|\, \hat{A}u^k(x)\|_{H^1(\Omega)}\big) \leqq \varepsilon c_3\,, \end{aligned} \tag{21}$$

where c_3 is a constant independent of ε. Here we used the fact that

$$\varrho\left(\frac{x}{\varepsilon}\right) - \hat{\varrho} = \varepsilon \frac{\partial}{\partial x_i} g^i\left(\frac{x}{\varepsilon}\right),$$

where $g^i(\xi)$ are bounded 1-periodic in ξ functions, and the fact that

$$\|u^{\varepsilon,k}(x)\|_{H^1(\Omega)} \leqq c_4|\lambda_k^\varepsilon| \leqq c_5(k)\,,$$

where $c_5(k)$ does not depend on ε. Estimate $|\mu_k(\varepsilon) - \mu_k| \leqq c(k)\sqrt{\varepsilon}$ follows from inequalities (19)—(21). The theorem is proved.

References

1 DE GIORGI, E.; SPAGNOLO, S., Sulla convergenza degli integrali dell'energia per operatori ellittici del secondo ordine. Boll. Un. Mat. Ital., (4), 8 (1973), 391—411.
2 ZHIKOV, V. V.; KOZLOV, S. M.; OLEINIK, O. A.; HA TIEN NGOAN, Homogenization and G-convergence of differential operators. Uspehi Mat. Nauk., 1979, 34:5, p. 63—133.
3 ZHIKOV, V. V.; OLEINIK, O. A., On homogenization of system of elasticity with almost periodic coefficients. Vestnik. Mosc. Univ. ser. 1, mat. mech., 1982 ,no. 6, p. 62—70.
4 YOSIFIAN, G. A.; OLEINIK, O. A.; SHAMAEV, A. S., Homogenization of eigenvalues and eigenfunctions of the boundary value problem of elasticity in a perforated domain. Vestn. Mosc. Univ., ser. 1, mat., mech., 1983, no. 4, p. 53—63.
5 YOSIFIAN, G. A.; OLEINIK, O. A.; SHAMAEV, A. S., Homogenization of eigenvalues of the boundary value problem of elasticity with rapidly oscillating periodic coefficients. Sibirsk. Matem. Zh. 1983, no. 5.
6 SHAMAEV, A. S., Homogenization of solutions and eigenvalues of boundary value problems for elliptic equations in perforated domains. Uspehi. Mat. Nauk., 1982, 37:2, p. 243—244.
7 OLEINIK, O. A.; YOSIFIAN, G. A.; PANASENKO, G. P., Asymptotic expansion of solutions of the elasticity system with rapidly oscillating coefficients. Dokl. Akad. Nauk SSSR, 1982, 266:1, p. 18—22.
8 OLEINIK, O. A.; YOSIFIAN, G. A.; PANASENKO, G. P., Asymptotic expansion of solutions of the elasticity system in perforated domains. Matem. Sbornik, 1983, 120:1, p. 22—41.
9 BENSOUSSAN, A.; LIONS, J. L.; PAPANICOLAOU, G., Asymptotic analysis for periodic structures., North Holland Publ. Co., 1978.
10 SANCHEZ-PALENCIA, E., Non-homogeneous media and vibration theory., Lect. Notes in Physics, Springer Verlag, 1980, 127.
11 OLEINIK, O. A.; YOSIFIAN, G. A., Estimate of the difference between the solution of the elasticity system in a perforated domain and the solution of the homogenized system, Uspehi Mat. Nauk, 1982, 37:5, p. 195—196.
12 VAINIKKO, G. M., Regular convergence of operators and approximate solution of equations. VINITI, Itogi Nauki i Techniki, ser. "Math. analysis" v. 16, 1979.

Received August 8, 1983

Address: Проф. О. А. Олейник, 117234 Москва, Ленинские Горы, МГУ, механико-математический факультет, СССР

BOOK REVIEWS

Lanckau, E. / Tutschke, W. (eds.), Complex Analysis. Methods Trends, and Applications. Berlin, Akademie-Verlag 1983. 298 S., 28 Abb., M 68,–. BN 763 109 6 (6699)

Dieses hochklassige Buch ist aus mehreren Tagungen hervorgegangen, die an der Universität Halle-Wittenberg stattgefunden haben. Sein Ziel ist, durch Übersichtsartikel (meist ohne Beweis) die Kommunikation zwischen der Funktionentheorie und Anwendungsgebieten zu fördern. Dabei werden Verallgemeinerungen der holomorphen Funktionen zugelassen, und bei den Anwendungen stehen gewöhnliche und partielle Differentialgleichungen, aber auch Integralgleichungen und Wahrscheinlichkeitstheorie im Vordergrund. Die Artikel sind in sich abgeschlossen, geben die wichtigsten Resultate an und sind ausgiebig mit Literaturhinweisen versehen; das gemeinsame Literaturverzeichnis umfaßt 48 Seiten. Die Liste der Autoren bürgt für hohe Qualität des Inhalts, und auch die Aufmachung läßt keine Wünsche offen. — Es folgt eine Übersicht über die 22 Kapitel des Buches: W. Tutschke: Classical and modern methods of complex analysis. H. Begehr: Value distribution theory. M. B. Balk: Polyanalytic functions. O. Forster: Cohomological methods in complex analysis. B. Bojarski: Connections between complex and global analysis: Riemann-Hilbert transmission problem. J. Ławrynowicz and G. Tsagas: On some complex-analytic methods in differential geometry. Z. D. Usmanov: Complex methods in the theory of infinitesimal bending of a surface. S. Prössdorf: Approximation methods for solving singular integral equations. E. Mues: Ordinary differential equations in the complex domain. H. Begehr: Boundary value problems for analytic and generalized analytic functions. V. S. Vinogradov: Elliptic systems in the plane. R. P. Gilbert: Recent results and developments in generalized hyperanalytic function theory. A. Džuraev: Complex methods in the theory of boundary value problems. H. Florian and R. Heersink: Differential operators in the theory of elliptic equations and boundary value problems. E. Lanckau: Solving of linear partial differential equations by using complex integral transforms. K. Habetha: Function theory in algebras. D. L. Colton: Analytic continuation of solutions to partial differential equations with applications to problems in scattering theory. W. L. Wendland: Numerical methods for boundary value problems of elliptic systems in the plane. L. v. Wolfersdorf: Optimal control problems with elliptic differential equation systems of first order in the plane. G. F. Mandžavidze: Methods of the theory of analytic functions in the theory of elasticity. E. I. Obolašvili: Some applications of generalized functions in the shell theory. H.-J. Rossberg and B. Jesiak: Complex analysis in probability theory.

Gießen D. Gaier

Alexits, G., Approximation Theory. Selected Papers. Ed. by Tandori, K., Budapest, Akadémiai Kiadó 1983. 297 S., Ft 330,–. ISBN 963-05-3204-2

George Alexits ist einer der hervorragendsten Vertreter der ungarischen Mathematik unseres Jahrhunderts, der neben seinem reichhaltigen mathematischen Schaffen eine aktive Rolle im öffentlichen Leben der Ungarischen Volksrepublik gespielt hat. Der vorliegende Band beinhaltet eine Sammlung von 34 Originalarbeiten von G. Alexits zur Approximationstheorie, die seinem Wirken vom Ende der zwanziger Jahre bis kurz vor seinem Tod im Jahr 1978 entstammen. An Hand dieser Arbeiten läßt sich die stürmische Entwicklung der Approximationstheorie gewissermaßen am Lebenswerk eines ihrer erfolgreichsten Vertreter nachempfinden. Die in diesem Sammelband aufgenommenen Zeitschriftenartikel sind in der Originalsprache wiedergegeben, lediglich für die original ungarisch abgefaßten Arbeiten wurde eine englische Übersetzung besorgt.

Der Sammelband wird durch eine mehrseitige Lebensdarstellung dieses bedeutenden Mathematikers und hervorragenden Lehrers und ein Verzeichnis aller seiner Veröffentlichungen eingeleitet. Der Sammelband wird durch eine kurze Wertung und Einordnung des Gesamtwerks von G. Alexits abgeschlossen.

Karl-Marx-Stadt F. Kuhnert

Pazy, A., Semigroups of Linear Operators and Applications to Partial Differential Equations. Berlin-Heidelberg-New York-Tokyo, Springer-Verlag 1983. VIII, 279 S., DM 88,–. US $ 34.20. ISBN 3-540-90845-5 (Applied Mathematical Sciences 44)

Das Interesse an Halbgruppen von Operatoren beruht in starkem Maße auf deren enger Beziehung zu Anfangswertaufgaben für Evolutionsgleichungen. Der Verfasser beschränkt sich in seinem Buch, das eine erweiterte Fassung von Vorlesungsausarbeitungen darstellt, auf Halbgruppen beschränkter linearer Operatoren in einem Banach-Raum. Dies ermöglicht eine einfache und in sich abgeschlossene Darlegung der Theorie, die vom Leser nur elementare Kenntnisse der Funktionalanalysis verlangt.

In Kapitel 1 wird der grundlegende Begriff des erzeugenden Operators einer Halbgruppe eingeführt, und es werden notwendige und hinreichende Bedingungen dafür angegeben, daß ein Operator eine Halbgruppe erzeugt (Sätze von Hille-Yosida und Lumer-Phillips). In Kapitel 2 werden Eigenschaften von Halbgruppen, wie Stetigkeit, Kontraktivität, Differenzierbarkeit und Analytizität, diskutiert und mit Eigenschaften der Resolvente des erzeugenden Operators in Verbindung gebracht. Kapitel 3 befaßt sich mit der Approximation von Halbgruppen und mit Störungen von erzeugenden Operatoren. In den Kapiteln 4 und 5 werden Anfangswertaufgaben für Evolutionsgleichungen mit zeitunabhängigen bzw. zeitabhängigen linearen Operatoren systematisch mit Hilfe der Theorie der Halbgruppen untersucht. Dabei erweist sich eine Unterteilung in „hyperbolische" und „parabolische" Probleme als zweckmäßig. Kapitel 6 zeigt, wie Informationen über Halbgruppen linearer Operatoren zur Lösung nichtlinearer Evolutionsgleichungen herangezogen werden können. Die beiden letzten Kapitel enthalten Anwendungen der abstrakten Theorie auf konkrete Rand-Anfangswertprobleme für lineare und nichtlineare Differentialgleichungen. Die in diesem Zusammenhang benötigten Resultate für elliptische Randwertprobleme werden ohne Beweis zitiert. Behandelt werden unter anderem die Schrödinger-Gleichung, die Wellengleichung, Varianten der Wärmeleitungsgleichung und die Korteweg-de Vries-Gleichung. Die dargestellten Ergebnisse umfassen allerdings nur einen Teil dessen, was heute über diese Gleichungen bekannt ist. Dies ist dadurch bedingt, daß sich der Verfasser bewußt auf die Anwendung der Theorie der Halbgruppen beschränkt hat. Hinweise auf Literatur, in der auch andere Zugänge zu Evolutionsgleichungen behandelt werden, findet der Leser in den am Schluß des Buches zusammengestellten bibliographischen Anmerkungen.

Inhaltlich weist das Buch zwangsläufig viele Überschneidungen mit früheren Büchern zu ähnlichen Themen auf, beispielsweise mit S. G. Krein, Lineare Differentialgleichungen im Banach-Raum (Moskau 1967, russisch), und mit H. Tanabe, Equations of Evolution (London 1979). Dies schmälert aber in keiner Weise den Wert des vorliegenden Buches. Auf Grund seines klaren Aufbaus und seiner sorgfältigen Ausführung ist es sowohl zur selbständigen Einarbeitung in den behandelten Themenkreis als auch als Grundlage für Spezialvorlesungen gut geeignet.

Berlin K. Gröger

Gruber, P. M. / Wills, J. M. (eds.), Convexity and Its Applications. Basel-Boston-Stuttgart, Birkhäuser Verlag 1983. 421 S., sFr. 110.–. ISBN 3-7643-1384-6

Das Buch enthält 16 in sich abgeschlossene Artikel über verschiedenartige Aufgabenklassen und Theorien, in denen konvexe Mengen eine wesentliche Rolle spielen. Sein Inhalt ist entsprechend breit gefächert. Man findet historische Betrachtungen (Fenchel) und moderne Bearbeitungen klassischer geometrischer Probleme (z. B. Chakerian/Groemer: Konvexe Körper konstanter Breite), Bezüge zur Differentialgeometrie (Leichtweiss), algebraische Untersuchungen (z. B. Chalk: Algebraische Gitter), Beiträge zur Konvexität in Banachräumen (z. B. von Phelps), Anwendungen in der kombinatorischen Optimierung (Bachem) sowie in Naturwissenschaften (Weil: Stereometrie), einen langen Artikel über additive Bewertungen konvexer Mengen (McMullen/Schneider) und vieles andere mehr. Das Buch dürfte wegen seiner Vielfalt und Tiefgründigkeit einen breiten Interessentenkreis finden. Sicher freuen sich auch viele über die sehr umfangreichen Literaturangaben.

Leipzig H. Hartwig

ZAMM · Z. Angew. Math. u. Mech. 65 (1985) 1, 19 – 27

SCHULTRICH, B.

Modified Propagation of Ultrasonic Surface Waves by Thin Elastic or Viscous Films

Die mechanischen Eigenschaften dünner Schichten unterscheiden sich oft von denen des kompakten Materials und hängen von den Herstellungsbedingungen ab. Sie lassen sich über das Ausbreitungsverhalten von Ultraschall-Oberflächenwellen direkt im Schicht-Substrat-Verbund bestimmen. Dazu wird für im Vergleich zur Wellenlänge kleine Schichtdicken das Verschiebungsfeld in der Schicht entwickelt. Das für den allgemeinen Fall angegebene Gleichungssystem wird für elastische und viskose Deckschichten in niedrigster Ordnung von d/λ ausgewertet. Für einen elastischen Film ist die relative Geschwindigkeitsänderung proportional zu $\omega \cdot d$, wobei der Proportionalitätsfaktor die relativen Unterschiede der elastischen Konstanten und der Dichten von Schicht und Substrat enthält. Durch eine zähe Schicht wird die Oberflächenwelle proportional zu $\omega^3\eta\, d$ gedämpft. Der elastische Fall bestätigt in niedrigster Näherung im wesentlichen eine früher von Achenbach und Keshava angegebene Näherung, die auf der Plattentheorie beruhte. Die dort angegebenen Terme höherer Ordnung in d/λ erweisen sich jedoch im Lichte einer systematischen Entwicklung als unvollständig.

The mechanical properties of thin films often differ from those of bulk material and depend on technological parameters. But they can be determined in the coating-substrate composite directly by means of ultrasonic surface waves. For thin layers in relation to wavelength, the displacement field within the layer is expanded in powers d/λ. The corresponding equation system given for the general case is evaluated for elastic and viscous coatings in lowest order of d/λ. An elastic covering layer causes a relative change in velocity proportional to $\omega \cdot d$ where the factor is determined by the elastic and inertial differences between film and substrate. By a viscous coating the surface wave is attenuated proportional to $\omega^3\eta d$. The elastic case confirms in the first order the results of a plate theory approximation given formerly by Achenbach and Keshava. However, their higher order terms prove to be incomplete from the viewpoint of the present systematic expansion.

Механические свойства тонких слоев часто отличаются от свойств сплошного материала и зависят от условий их изготовления. Но они определимы непосредственно с помощью ультразвуковых поверхностных волн. Для тонких слоев перемещение внутри слоя разлагается в степенный ряд по d/λ. Общая система уравнений используется для упругих и вязких покрытий в первом приближении d/λ. Упругий слой вызывает изменение скорости пропорциональное $\omega \cdot d$, причём фактор пропорциональности зависит от различия модулей Юнга или плотностей в слое и в подслое. Вязким покрытием поверхностная волна ослабляется по закону $\omega^3 \cdot \eta \cdot d$. Упругий случай подтверждает в первом приближении результаты Ахенбаха и Кешавы, достигнутые апроксимацией пластины. Но их члены высших порядков являются неполными с точки зрения систематического разложения.

1. Introduction

Acoustical surface waves (Rayleigh waves) are travelling along the surface of an elastic, homogeneous and extended body without any dispersion caused by geometry [1]:

$$c_R = \frac{\omega}{k} = \frac{1}{s(\nu)}\, c_t \qquad \text{with} \quad \Phi(\omega, k) = (k^2 + \varkappa_t^2)^2 - 4k^2\varkappa_l\varkappa_t = 0 \tag{1}$$

or

$$(2s^2 - 1)^2 - 4s^2\sqrt{s^2 - 1}\,\sqrt{s^2 - \frac{1 - 2\nu}{2 - 2\nu}} = 0\,, \qquad s(\nu) \approx \frac{1 + \nu}{0.87 + 1.12\nu}\,.$$

This is evident from dimensional reasons because neither internal nor external characteristic lengths exist to which the wave length can be related. But such a length is available in media with a stratified surface region, the wave propagation depends on wavelength and frequency now. For the first time, such problems arose from investigations of seismic waves. In this field the layer thicknesses d usually exceed the wavelength: $d \cdot k \gtrsim 1$. Therefore theoretical evaluation was concentrated on this part of dispersion relation $k(\omega)$, which can only be numerically described, and especially on the principal changes in mode structure: For sufficient short waves and in the case of total reflection at the boundary ($c_t < c_t'$) transversal Love waves originate from loss-free zig-zag-reflections in the layer, just in analogy to the electromagnetic field in waveguides [2]. Besides a single solid, elastic coating [3], fluid, volume-elastic, non-viscous layers [4], multilayers [4] and materials with continuously varying elastic properties [5] are considered, too. In the application of surface waves as a nondestructive probe for materials testing, the interesting thicknesses are also of the order of wavelength $d \cdot k \approx 1$: The measurement of thickness or elastic properties of coatings by means of the changed velocity of ultrasonic surface waves ($\lambda \approx 0.05 \dots 1$ mm) are mainly carried out at galvanic and similar metallic layers with $d \gtrsim 0.1$ mm [6]. A new situation arises for thin films below 0.01 mm as they are used in microelectronics and as protective coatings against wear and corrosion. Here the film thickness usually is much smaller then the applied wave length: $d \cdot k \ll 1$. Therefore, in this paper the modification of dispersion relation caused by thin-film coating is evaluated in the lowest order of $\delta \equiv d \cdot k \approx d/\lambda$. The simple analytic expressions for changing the wave velocity and damping, following in this way, make allowance for direct determination of elastic and viscous properties of the film on substrate. Measurements by means of static methods are possible for free layers or films on thin substrate foils, but producible in special cases, or for bulk material, only. However the properties of thin films are highly dependent on technological conditions and interaction with the substrate during the deposition process and therefore can markedly differ from those of model systems.

The nature of these relations already can be derived from simple estimations. Therefore in the elastic case the heterogenous material is replaced by an effective homogeneous one with changed elastic modulus and density by considering the penetration depth of the order of wave length λ:

$$E + \Delta E \approx E + \frac{d}{\lambda}(E' - E), \qquad \varrho + \Delta\varrho \approx \varrho + \frac{d}{\lambda}(\varrho' - \varrho). \tag{2}$$

With $c \approx \sqrt{E/\varrho}$ we get for the relative shift of surface wave velocity

$$\frac{\Delta c}{c} \approx \frac{1}{2}\left(\frac{\Delta E}{E} - \frac{\Delta\varrho}{\varrho}\right) \approx \frac{1}{2}\frac{d}{\lambda}\left(\frac{E' - E}{E} - \frac{\varrho' - \varrho}{\varrho}\right) \tag{3}$$

(quantities with a prime concern the coating).
The damping in the case of viscous coatings results from the proportion of dissipated energy to stored energy. The dissipated energy $\Delta U(l)$ can be estimated as the product of power density $\eta'\dot{\varepsilon}^2 \approx \eta'\omega^2\varepsilon^2$, of travelled specimen volume $l \cdot B \cdot d$ (l = travelled distance, B = specimen width) and of travelling time l/c. The stored energy U is approximately the product of energy density in the elastic substrate $E\varepsilon^2$ and of volume excited by the surface wave $\lambda \cdot l \cdot B$. From this the attenuation factor α, the relative energy dissipation by length, follows:

$$\alpha \equiv \frac{1}{l}\,\Delta U(l)/U \approx \frac{d}{\lambda}\frac{\eta'\omega^2}{Ec} = \frac{1}{2\pi}\frac{d\eta'}{E}\frac{\omega^3}{c^2}. \tag{4}$$

An explicit expression for shifting the phase velocity of acoustic surface waves by an elastic coating can be gathered from an approximated solution given by Achenbach and Keshava [7]. They simplified the displacement field within the surface layer according to Mindlin's plate equations. These are coupled with the substrate by means of continuity conditions of displacements and stresses along the boundary. Hence it follows the dispersion relation $k(\omega)$ (equation (48) in [7]) in the form

$$0 = a_4(\omega, k)(kd)^4 + a_3(\omega, k)(kd)^3 + a_2(\omega, k)(kd)^2 + a_1(\omega, k)(kd) + a_0(\omega, k). \tag{5}$$

In the lowest order of $k \cdot d$, the sufficient approximation for thin films, we get

$$a_0(\omega, k) = -(kd)\, a_1(\omega, k) \tag{6}$$

with

$$\left.\begin{aligned} a_0 &= (k^2 + \varkappa_t^2)^2 - 4k^2\varkappa_t\varkappa_l = \Phi(\omega, k), \\ -a_1 &= 4\frac{G'}{G}\frac{\gamma'^2 - 1}{\gamma'^2}\varkappa_t k(k^2 - \varkappa_t^2) - \frac{\varrho'}{\varrho}(k^2 - \varkappa_t^2)^2\frac{\varkappa_t + \varkappa_l}{k}. \end{aligned}\right\} \tag{7}$$

For disappearing coating ($d \to 0$) the usual condition for Rayleigh waves $\Phi = 0$ without any frequency dependence follows. This also has to apply if the coating material is similar to the substrate, because the separation of substrate and layer looses its physical importance, then. For the first sight this is not evident in (6) and this fact was the starting point for the present revived evaluation. Moreover it was not clear if and how the equations of Achenbach and Keshava could be transformed for other linear constitutive equations of the layer, especially for viscous material behaviour. Therefore in the following a systematic expansion of layer-displacement field in powers of $\delta = k \cdot d$ is given. At the same time this clear approach will throw new light upon the validity of the nonlinear terms in (5).

2. Propagation of boundary waves within the substrate

The elastic behaviour of the substrate is determined by the equations of motion

$$\varrho\frac{\partial^2 u_i}{\partial t^2} = \sigma_{ik,k} \tag{8}$$

and the constitutive equation

$$\sigma_{ik} = G(u_{i,k} + u_{k,i} + (\gamma^2 - 2)\,\delta_{ik}u_{l,l}). \tag{9}$$

Both systems can be condensed to

$$\frac{\partial^2\vec{u}}{\partial t^2} = -c_t^2 \operatorname{curl}\operatorname{curl}\vec{u} + c_l^2 \operatorname{grad}\operatorname{div}\vec{u}. \tag{10}$$

The general solution is representable as superposition of a transversal, i.e. divergence-free part and a longitudinal, i.e. curl-free part:

$$\vec{u} = \vec{u}_t + \vec{u}_l \quad \text{with} \quad \operatorname{div}\vec{u}_t = 0 \quad \text{and} \quad \operatorname{curl}\vec{u}_l = 0. \tag{11}$$

Thus (10) is divided into two disconnected wave equations. A wave, propagating along x-direction in the surface region of a body, which is filling half space $z \leqq 0$, has the general form

$$\vec{u}_t = \vec{a}\,e^{i(kx-\omega t)}\,e^{\varkappa_t z}, \qquad \vec{u}_l = \vec{b}\,e^{i(kx-\omega t)}\,e^{\varkappa_l z}. \tag{12}$$

The wave equations (and with them the equations of motion) are identically fulfilled if the relations

$$\varkappa_t^2 = k^2 - \frac{\omega^2}{c_t^2}, \quad \varkappa_l^2 = k^2 - \frac{\omega^2}{c_l^2} \tag{13}$$

and if the curl and divergence conditions

$$a_x = \varkappa_t a\,, \quad a_z = -i \cdot k \cdot a\,, \quad b_x = k \cdot b\,, \quad b_z = -i\varkappa_l b \tag{14}$$

hold true. The dispersion relation $k(\omega)$ follows from the solubility condition for the remaining unknown quantities a and b, which in their turn are determined by the continuity conditions for the stresses at boundary $z = 0$

$$\left.\begin{aligned} \sigma_{zz}(z = 0) &= -iG[2k\varkappa_t a + (k^2 + \varkappa_t^2)\, b]\, e^{i(kx-\omega t)}\,, \\ \sigma_{xz}(z = 0) &= G[(\varkappa_t^2 + k^2)\, a + 2k\varkappa_l b]\, e^{i(kx-\omega t)}\,. \end{aligned}\right\} \tag{15}$$

For the case of a free surface with $\sigma_{zz} = 0$ and $\sigma_{xz} = 0$ this yields the condition $\Phi = 0$, cited above, for usual Rayleigh-waves with constant velocity c_R according to (1). However, if substrate is covered by another material of thickness d, then the stresses and displacements at boundary are influenced by the properties of the coating and thus the dispersion relation is modified.

3. Propagation for surface waves in coated materials

In addition to the displacement field (11) and (12) within the substrate an analogous ansatz has to be assumed for the coating, too. Because of its finiteness the exponentially increasing terms cannot be excluded now

$$\vec{u} = \{\vec{a}'_+ e^{\varkappa_i z} + \vec{a}'_- e^{-\varkappa_i z} + \vec{b}'_+ e^{\varkappa_i z} + \vec{b}'_- e^{-\varkappa_l z}\}\, e^{i(kx-\omega t)} \qquad (0 \leqq z \leqq d)\,. \tag{16}$$

By means of continuity conditions for displacements u_i and stresses σ_{iz} at the boundary the dispersion relation follows as a vanishing determinant of the 6th order [2, 6]. According to our limited problem, the wave propagationat thin films, we concentrate only upon the lowest terms in $\varkappa' \cdot z$ and replace (16) by

$$\vec{u}' = (\vec{A} + \vec{B}(kz) + \vec{C}(kz)^2)\, e^{i(kx-\omega t)} \qquad (0 \leqq z \leqq d)\,. \tag{17}$$

This approximation applies as long as

$$|\varkappa_s' d| = |\sqrt{1 - (c/c_s')^2}\, k \cdot d| \ll 1 \qquad (s = t, l)\,. \tag{18}$$

Thus, because of $c \approx c_R \approx c_t$ the film must be thin compared with the smaller one of transversal wavelengths of substrate and coating:

$$\left.\begin{aligned} &k \cdot d \ll 1 \quad \text{or} \quad d \ll \lambda_t/2 \qquad (c_t \lesseqgtr c_t')\,, \\ &k \cdot dc_t/c_t' \ll 1 \quad \text{or} \quad d \ll \lambda_t'/2 \qquad (c_t \gtreqless c_t')\,. \end{aligned}\right\} \tag{19}$$

(For abbreviation in the following we speak about small quantities $\delta = k \cdot d$ only independent on discrimination (19)). Continuity of displacements at boundary $z = 0$ yields

$$\vec{A} = \vec{u}(z = 0) = \vec{a} + \vec{b}\,. \tag{20}$$

Furthermore the normal stresses at free surface $z = d$ must vanish and those at boundary $z = 0$ have to be continuous:

$$\sigma'_{xz}(z = d) = 0\,, \quad \sigma'_{zz}(z = d) = 0\,, \quad \sigma_{xz}(z = 0) = \sigma'_{xz}(z = 0)\,, \quad \sigma_{zz}(z = 0) = \sigma'_{zz}(z = 0)\,. \tag{21}$$

They are completed by the equations of motion within the coating, whereas in the substrate these equations are fulfilled identically by means of the separation ansatz (12). According to approximation (16), which neglects refinements of deformation field in the thin film, we concentrate to the satisfaction of the section averaged equations. From

$$\int \mathrm{d}V \varrho \frac{\partial^2 u_i}{\partial t^2} = \oint\!\!\oint \mathrm{d}A_k \sigma_{ik}$$

we get by evaluation for a small slice $\mathrm{d}x$ of the coating:

$$\left.\begin{aligned} -\omega^2 d\varrho' \bar{u}_x &= ikd\overline{\sigma_{xx}} - \sigma_{xz}(0) + \sigma_{xz}(d) = ikd\overline{\sigma_{xx}} - \sigma_{xz}(0)\,, \\ -\omega^2 d\varrho' \bar{u}_z &= ikd\overline{\sigma_{xz}} - \sigma_{zz}(0) + \sigma_{zz}(d) = ikd\overline{\sigma_{xz}} - \sigma_{zz}(0)\,. \end{aligned}\right\} \tag{22}$$

(Herein $\bar{u}_i$ and $\overline{\sigma_{ik}}$ denote quantities averaged across layer section:

$$\bar{u}_i = \frac{1}{d}\int_0^d \mathrm{d}z\, u_i(z)\,, \qquad \overline{\sigma_{ik}} = \frac{1}{d}\int_0^d \mathrm{d}z\, \sigma_{ik}(z)$$

For an elastic coating with the constitutive equation

$$\sigma'_{ik} = G'(u'_{i,k} + u'_{k,i} + (\gamma'^2 - 2)\, \delta_{ik} u'_{l,l}) \tag{23}$$

equations (20) to (23) can be transformed into a linear system for coefficients A_i, B_i and C_i ($i = x, z$):

$$\left.\begin{aligned} 0 &= \delta B_x + 2\delta^2 C_x + i\delta(A_z + \delta B_z + \delta^2 C_z)\,, \\ 0 &= i\delta(\gamma'^2 - 2)\,(A_x + \delta B_x + \delta^2 C_x) + \delta\gamma'^2(B_z + 2\delta C_z)\,, \end{aligned}\right\} \tag{24}$$

$$\left.\begin{aligned} G\{(k^2 + \varkappa_t^2)\,a + 2k\varkappa_l b\} &= G'k(B_x + iA_z)\,, \\ -iG\{2k\varkappa_t a + (k^2 + \varkappa_t^2)\,b\} &= G'k\{i(\gamma'^2 - 2)\,A_x + \gamma'^2 B_z\}\,, \end{aligned}\right\} \tag{25}$$

$$\left.\begin{aligned} &-\omega^2 d\varrho'\,(A_x + \tfrac{1}{2}\delta B_x + \tfrac{1}{3}\delta^2 C_x) = \\ &= ik\delta G'\{\gamma'^2 i(A_x + \tfrac{1}{2}\delta B_x + \tfrac{1}{3}\delta^2 C_x) + (\gamma'^2 - 2)\,(B_z + \delta C_z)\} - G'k(B_x + iA_z)\,, \\ &-\omega^2 d\varrho'(A_z + \tfrac{1}{2}\delta B_z + \tfrac{1}{3}\delta^2 C_z) = \\ &= ik\delta G'\{B_x + \delta C_x + i(A_z + \tfrac{1}{2}\delta B_z + \tfrac{1}{3}\delta^2 C_z\} - G'k\{i(\gamma'^2 - 2)\,A_x + \gamma'^2 B_z\}\,. \end{aligned}\right\} \tag{26}$$

In lowest order of $\delta = k \cdot d$ equations (25) and (26) already give a closed system for determination of A_i and B_i. The coefficients C_i of the second order terms in (17) follow from (24), then. (The conditions for free surface $\sigma'_{iz}(z = d) = 0$ are taken into consideration via (22) and (26).)

The equations for A_i and B_i are solvable in a nontrivial manner if and only if the coefficient determinant vanishes. Thus we get the dispersion relation $k(\omega)$ in first order of δ (see Appendix):

$$\Phi(\omega, k) = -\delta\,\frac{\varrho'}{\varrho}\,(k^2 - \varkappa_t^2)^2\,\frac{\varkappa_t + \varkappa_l}{k} + 4\delta\,\frac{G'}{G}\,\frac{\gamma'^2 - 1}{\gamma'^2}\,\varkappa_t k(k^2 - \varkappa_t^2)\,. \tag{27}$$

By making allowance for identity

$$\Phi(\omega, k)\,(\varkappa_t + \varkappa_l) = (k^2 - \varkappa_t^2)^2\,(\varkappa_t + \varkappa_l) - 4k^2\varkappa_t(k^2 - \varkappa_t^2)\,(1 - 1/\gamma^2) \tag{28}$$

it can be transformed to

$$\Phi(\omega, k)\,\big(1 + d(\varkappa_t + \varkappa_l)\big) = -\delta\,\frac{\varrho' - \varrho}{\varrho}\,(k^2 - \varkappa_t^2)^2\,\frac{\varkappa_t + \varkappa_l}{k} + 4\delta\,\frac{1}{G}\left(G'\,\frac{\gamma'^2 - 1}{\gamma'^2} - G\,\frac{\gamma^2 - 1}{\gamma^2}\right)\varkappa_t k(k^2 - \varkappa_t^2). \tag{29}$$

From (27) and (29) it is evident that all the limiting cases of vanishing film thickness ($d \to 0$), of coatings with negligible elastic and inertial effects ($G' \to 0$, $\varrho' \to 0$) and of layers, indiscernible from substrate all lead to the dispersion relation $\Phi(\omega, k) = 0$ of uncoated material.

Because of small perturbation only of wave propagation by thin films, the left-hand side can be expanded around the undisturbed wavenumber $k_0(\omega)$:

$$\Phi(\omega, k)\left(1 + \delta\,\frac{\varkappa_t + \varkappa_l}{k}\right) \approx \left.\frac{\mathrm{d}\Phi}{\mathrm{d}k}\right|_{k_0}(k - k_0) =$$

$$= 2(k - k_0)\,k_0\left\{4\left(2k_0^2 - \frac{\omega^2}{c_t^2}\right) - \frac{1}{k_0^2}\left(2k_0^2 - \frac{\omega^2}{c_t^2}\right)^2 - 8k_0^4\,\frac{2k_0^2 - \dfrac{\omega^2}{c_t^2} - \dfrac{\omega^2}{c_l^2}}{(2k_0^2 - \omega^2/c_t^2)^2}\right\} =$$

$$= -2\,\frac{k - k_0}{k_0}\left(\frac{\omega}{c_t}\right)^4 \sqrt{s^2 - 1}\,\frac{2\pi s^2}{1 - \nu}\,\frac{1}{p(\nu)}\,.$$

In the right-hand term proportional to δ, it is sufficient to replace the unknown quantity by the undisturbed k_0. Thus a relative change of wavenumber due to a thin coating is obtained:

$$\frac{k - k_0}{k_0} = -\frac{p(\nu)}{2\pi s(\nu)}\,k_0 d\left\{\frac{1 - \nu}{2G}\left(\frac{2G'}{1 - \nu'} - \frac{2G}{1 - \nu}\right) - q(\nu)\,\frac{\varrho' - \varrho}{\varrho}\right\}. \tag{30}$$

The auxiliary quantities

$$\begin{aligned} p(\nu) &\equiv 2\pi s^2\sqrt{s^2 - 1}\,\frac{1}{1 - \nu}\left\{8s^6\,\frac{2s^2 - 1 - 1/\gamma^2}{(2s^2 - 1)^2} + (2s^2 - 1)^2 - 4s^2(2s^2 - 1)\right\}^{-1} \\ &= \frac{\pi}{2}\,\frac{1}{1 - \nu}\sqrt{s^2 - 1}\,\frac{(2s^2 - 1)^2}{\dfrac{3}{1 - \nu}\,s^4 - 2\,\dfrac{2 - \nu}{1 - \nu}\,s^2 + 1} \end{aligned} \tag{31}$$

and

$$q(\nu) \equiv \frac{1 - \nu}{2}\,\frac{1}{s^2}\,\big(1 + \sqrt{(s^2 - 1/\gamma^2)/(s^2 - 1)}\big) \approx 1$$

with $s(\nu)$ according to (1) depend on Poisson modulus of substrate only (Fig. 1).

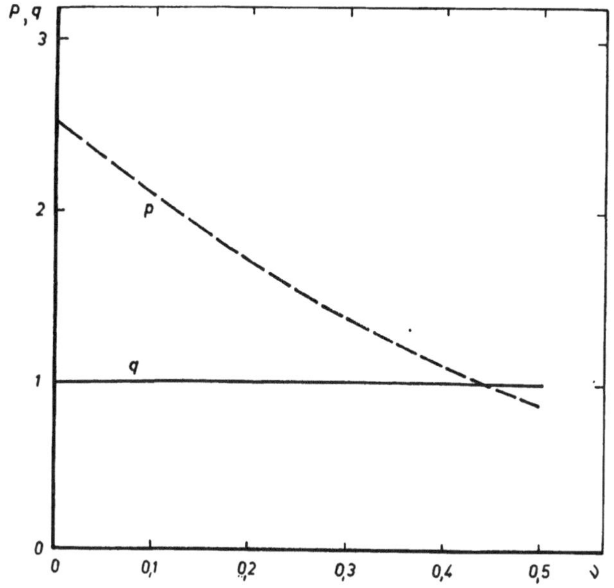

Fig. 1. Auxiliary functions $p(\nu)$ and $q(\nu)$ for the shift of wave velocity by an elastic coating

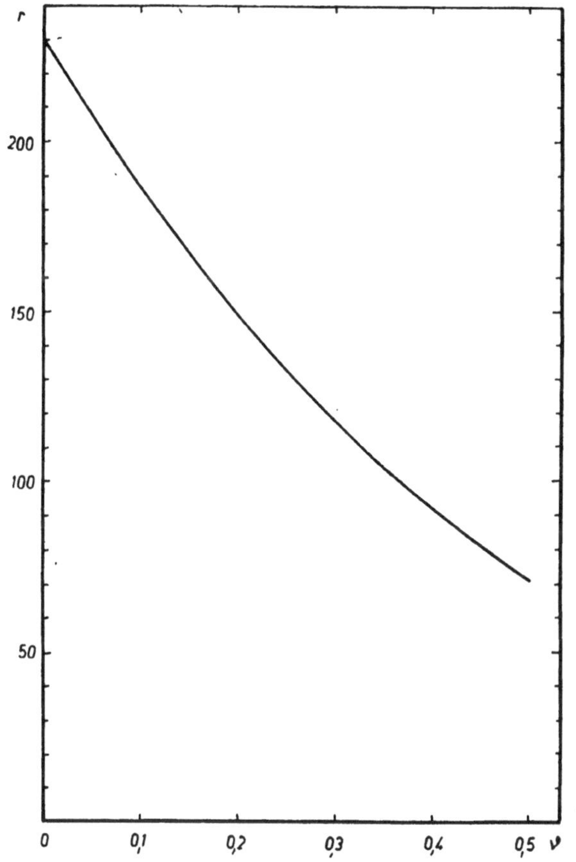

Fig. 2. Auxiliary function $r(\nu)$ for the attenuation of the surface wave by a viscous coating

4. Discussion: Elastic coating

If the substrate is coated by a material which behaves completely elastic, then G' and followingly k are real. The only effect of surface layer existis in a variation of phase velocity $c = \omega/k$ and group velocity $c_G = \mathrm{d}\omega/\mathrm{d}k$. Unlike the undisturbed surface wave they differ due to dispersion:

$$\left.\begin{aligned} c &= \frac{\omega}{k} = c_R\left(1 - \frac{k - k_0}{k_0}\right), & \frac{c - c_R}{c_R} &= -\frac{k - k_0}{k_0}, \\ c_G &= \frac{\mathrm{d}\omega}{\mathrm{d}k} = c_R\left(1 - 2\frac{k - k_0}{k_0}\right), & \frac{c_G - c_R}{c_R} &= -2\frac{k - k_0}{k_0}. \end{aligned}\right\} \tag{32}$$

In usual ultrasonic experiments the transit time of a wave packet and thus the group velocity is measured. Neglecting the slight difference between $1 - \nu'^2$ and $1 - \nu^2$ (as a rule below 10%) such measurements immediately can be connected with elastic modulus $E' = 2(1 + \nu')\, G'$:

$$\frac{c_G - c_R}{c_R} = 2p(\nu) \frac{d}{\lambda_t} \left\{ \frac{E' - E}{E} - q(\nu) \frac{\varrho' - \varrho}{\varrho} \right\}. \tag{33}$$

On account of $q(\nu) \approx 1$ this coincides with the rough estimation given in (3). The calculations leading to (33) are based on the assumption (19) $\delta = k \cdot d \ll 1$. However comparison with special cases of the numerically solved exact dispersion relation [2, 6] and with experimental results (Fig. 3) shows its validity up to $\delta = k \cdot d \lessapprox 0.6$, that means $d \lessapprox 0.1\lambda_t$.

Equation (33) is the basis for determining the elastic modulus of thin films by means of ultrasonic surface waves [8]. As an example Fig. 3 shows the increase of the group velocity of the surface waves on a cemented carbide specimen coated with a TiC layer. The velocities are obtained by measuring the time that a wave packet of surface waves needs to run over a given distance. For this a broad band impulse was produced by a laser flash, the interesting frequencies were selected from the arriving signal by an interdigital transducer. The data confirm the linear relationship between velocity shift and film thickness up to $d \lessapprox 0.1\lambda_t$. Calculation of Young's modulus of the TiC coating yields at 25 MHz and at 36 MHz accordingly the same value $E' = (373 \pm 10)$ GPa, which substantiates the frequency dependence anticipated in (33). This value lies below the elastic modulus of the substrate (493 GPa), but the lower density of the coating (4.9 g/cm³ compared with 11.0 g/cm³ of the cemented carbide) overcompensates the smaller stiffness and causes the coating material to have higher ultrasonic velocities than the substrate.

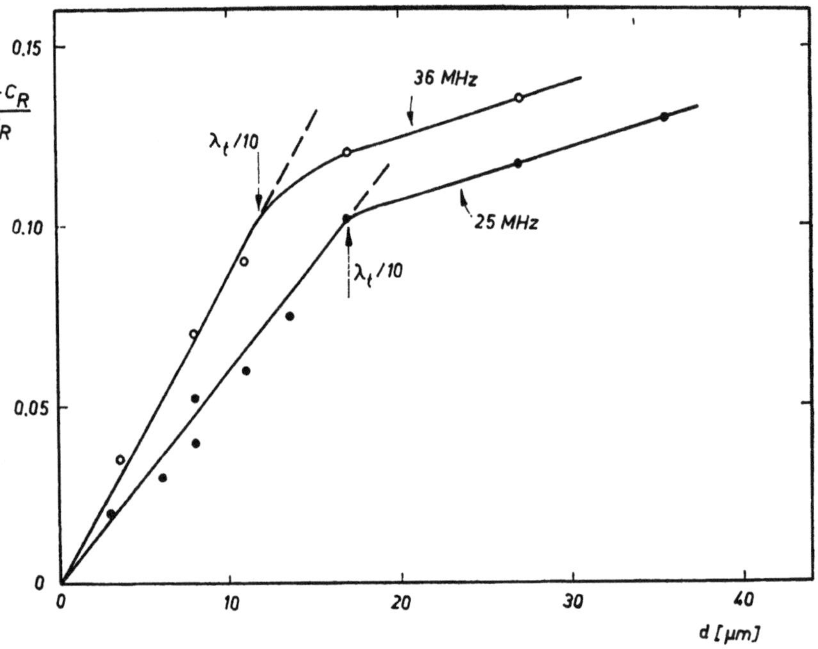

Fig. 3. Relative shift of surface wave group velocity of a TiC coated cemented carbide substrate versus film thickness d [8]. (Frequencies 25 MHz and 36 MHz)

By comparing (7) and (27) it can be seen that the plate theory approximation of Achenbach and Keshava (besides the unimportant difference $1/(1 - \nu^2)$ instead of $1/(1 - \nu'^2)$) yields the right first order correction in the dispersion relation. This coincidence follows from the analogue (but not identical) linear approximation of displacements within the layer. (In plate theory the stronger restriction $B_z = 0$ is made and the boundary stresses $\sigma_{zz}(z=0)$, $\sigma_{xz}(z = 0)$ are introduced as additional variables). However higher order terms in the dispersion relation require those of the same order in the displacements as it is evident from the system (24), (25), (26). Because lacking of such terms in their ansatz the approximated dispersion relation (5) misses terms of the same order and thus does not give the correct expansion. On contrary, the straightforward approach given above, which is based on the direct development of displacement field, can be extended systematically. For this the refinements of the field must be completed by a more rigorous consideration of the equations of motion. Therefore we fulfil (8) not only in the average taken over the layer section but for the moments, too:

$$-\omega^2 d\varrho' \overline{u_i z^l} = (\sigma_{iz} z^l)|_{z=d} - (\sigma_{iz} z^l)|_{z=0} + ikd\overline{\sigma_{ix} z^l} - ld\overline{\sigma_{iz} z^{l-1}} = ikd\overline{\sigma_{ix} z^l} - ld\overline{\sigma_{iz} z^{l-1}} \quad (l \geqq 1)\,. \tag{34}$$

The equations generalising (24) to (26) by including higher order terms are given in the Appendix. The condition for vanishing normal stress on free surfaces, which is analogous to (24), is not required for deriving the dispersion relation. It provides, however, next-order correction terms to the displacements, just like the second order terms C_i via (24) follow from the closed first order system (25), (26). This fact explains the empirical statement in [7] on large deviations of linear displacement field within the region where the linear approximation of dispersion relation is good.

5. Discussion: Viscous coating

For a substrate covered by a viscous fluid the material equation (23) has to be replaced by

$$\sigma'_{ik} = -p\delta_{ik} - i\omega\eta'(u'_{i,k} + u'_{k,i}) + (\beta' - \tfrac{2}{3}\eta')(-i\omega)\, u'_{l,l}\delta_{ik}\,. \tag{35}$$

Herein η' and $\beta' \equiv m\eta'$ denote the shear and bulk viscosity of the coating, respectively. The hydrostatic pressure p, built up in the fluid layer, is itself of order δ and causes higher order terms only. Thus it can be neglected within the framework of first-order approximation. Then the viscous material (35) can be formally treated as an elastic one (23) by replacing $G' \to -i\omega\eta'$ and $\gamma'^2 \to m + 4/3$. With this the dispersion relation (27) yields:

$$\Phi(\omega, k) = -\delta\,\frac{\varrho'}{\varrho}\,(\tilde{k}^2 - \tilde{\varkappa}_t^2)^2\,\frac{\tilde{\varkappa}_t + \tilde{\varkappa}_l}{\tilde{k}} - 4i\delta\,\frac{\omega\eta'}{G}\,\frac{3m+1}{3m+4}\,\tilde{\varkappa}_t\tilde{k}(\tilde{k}^2 - \tilde{\varkappa}_t^2)\,. \tag{36}$$

The correction of the wave number by coating now has an imaginary part $\tilde{k} = k + i\alpha/2$ (k real). The attenuation factor α describes the damping of the surface wave, the energy of which (proportional to the squared amplitude) decays along the distance l like $\exp(-\alpha l)$. Expanding (36) for small d/λ, we get the attenuation factor for thin viscous films:

$$\alpha = \frac{2}{\pi}\,\frac{3m+1}{3m+4}\,s(\nu)\,(1-\nu)\,p(\nu)\,\frac{\varrho\eta'}{G^2}\,d\omega^3\,. \tag{37}$$

Again the simple approximation given in the beginning is confirmed:

$$\alpha \approx \frac{p(\nu)}{\pi}\,\frac{d\eta'}{E}\,\frac{\omega^3}{c_t^2} \quad (\text{for } m \approx 0)\,.$$

By means of (37) the viscosity of thin coatings can be determined by measuring attenuation of ultrasonic surface waves [9]. For high viscous media and with the usual convention of vanishing bulk viscosity ($m = 0$), ultrasonic measurements by means of bulk waves yield approximately the same values for viscosity as rheological standard equipments [10]. The same should be valid for surface waves, too. Thus we get:

$$\alpha = r(\nu)\,\frac{df^3\eta'}{Ec_t^2}\,. \tag{38}$$

The auxiliary function

$$r(\nu) \equiv 8\pi^2 s(\nu)\,p(\nu) \tag{39}$$

is represented in Fig. 2.

In lowest order of d/λ the wave velocity is affected by inertia only:

$$\frac{c_G - c_R}{c_R} = -2p(\nu)\,\frac{\varrho'}{\varrho}\,\frac{df}{c_t}\,. \tag{40}$$

The velocity is diminished due to the additional oscillating mass without elasticity. Because density is not a structure sensitive quantity, the measurement of group velocity yields the film thickness which is needed for evaluation of viscosity from the attenuation experiment.

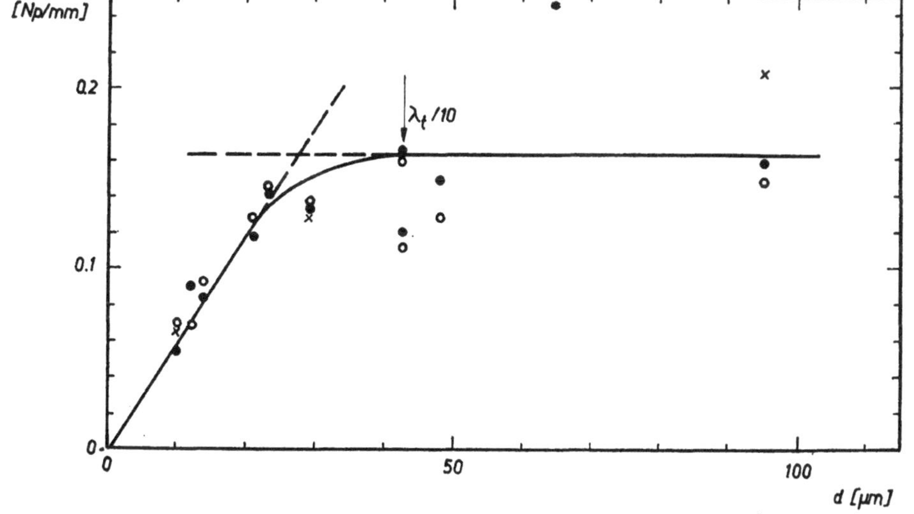

Fig. 4. Increase of the attenuation factor $\alpha_{\text{coated}} - \alpha_{\text{uncoated}}$ of a cemented carbide substrate with a visco-elastic calcium soap lubrication film versus film thickness d [9]. (Frequency 10 MHz, film length l = 10 mm (×), 20 mm (○) and 30 mm (●))

Fig. 4 shows the increase of the attenuation factor α of a cemented carbide substrate by a visco-elastic lubrication film on the base of calcium soaps at 10 MHz [9]. It was determined by measuring the decrease of the wave packet amplitudes for the coated substrate with different coating length compared to the uncoated substrate. A similar equipment as that mentioned above at Fig. 3 was used. The linear dependence of the attenuation factor on film thickness, which is characteristic for the proposed thin film approximation, is observed up to $d \approx 0.05\lambda_t$. According to (38) this corresponds to a film viscosity of $\eta' = 380$ Pa s. In this case of a solid visco-elastic soap coating it is more adequate to characterize the mechanical damping behaviour by the imaginary part $H' = 2\pi f\eta'$ of the complex elastic modulus [11]. For the soap film results a value of 24 GPa (compared with the static modulus of approximately 1.6 GPa).

References

1 Landau, L. D.; Lifschitz, E. M., Lehrbuch der theoretischen Physik, Bd. VII Elastizitätstheorie, Berlin 1966, p. 120.
2 Бреховских, Л. М., Волны в слоистых средах, Москва 1973.
3 Newlands, M., The disturbance due to a line source in a semiinfinite elastic medium with a single surface layer, Phil. Trans. Roy. Soc., Ser. A 245 (1952) 213.
4 Wedel, D., Dispersionserscheinungen elastischer Wellen in geschichteten Medien, Thesis, Bergakademie Clausthal 1964.
5 Kienczyłski, P., Propagation of surface SH waves in nonhomogeneous media, J. Tech. Phys., 22 (1981) 73.
6 Häusler, E., Elastische Wellen an der Oberfläche fester, insbesondere geschichteter Medien und ihre Anwendung auf die zerstörungsfreie Werkstoffprüfung, Materialprüfung, 2 (1960) 51.
7 Achenbach, J. D.; Keshava, S. P., Free waves in a plate supported by a semi-infinite continuum, J. Appl. Mechanics, 34 (1967) 397.
8 Schmidt, E., Untersuchung aufgetragener Hartstoffschichten mit laserinduzierten Ultraschall-Oberflächenwellen, report, Institut f. Physik d. Werkstoffbearbeitung, Berlin 1981.
9 Schneider, D.; Schultrich, B., Untersuchung des Relaxationsverhaltens stark dämpfender Schichten mittels Ultraschallflächenwellen, to be published in Experimentelle Tecxnik der Physik.
10 Bergmann, L., Der Ultraschall und seine Anwendung in Wissenschaft und Technik, Stuttgart 1954, p. 471.
11 Litovitz, T. A.; Davis, C. M., Structural and Shear Relaxation in Liquids, in W. P. Mason (Ed.), Physical Acoustics, Vol. II A, New York 1965, p. 282.

Received September 30, 1983, in original form January 3, 1983

Address: Dr. sc. nat. Bernd Schultrich, Zentralinstitut für Festkörperphysik und Werkstofforschung, 8027 Dresden, Helmholtzstr. 20, DDR

Appendix 1

Solution of the system (24)—(26)

By neglecting terms proportional to δ^2 and considering

$$\omega^2 \frac{d}{k} \frac{\varrho'}{G'} = \frac{\varrho'}{\varrho} \frac{G}{G'} \frac{d(k^2 - \varkappa_t^2)}{k} \approx \delta$$

equation (26) can be transformed to

$$\frac{\varrho'}{\varrho} \frac{G}{G'} \frac{d}{k} (k^2 - \varkappa_t^2) A_x = B_x + iA_z - i\delta \frac{\gamma'^2 - 2}{\gamma'^2} (i(\gamma'^2 - 2) A_x + \gamma'^2 B_z) + 4\delta \frac{\gamma'^2 - 1}{\gamma'^2} A_x$$

$$\frac{\varrho'}{\varrho} \frac{G}{G'} \frac{d}{k} (k^2 - \varkappa_t^2) A_z = i(\gamma'^2 - 2) A_x + \gamma'^2 B_z - i\delta(B_x + iA_z).$$

By means of (25) the coefficients B_l are eliminated:

$$H_1(a, b) - \delta \frac{\gamma'^2 - 2}{\gamma'^2} H_2(a, b) + 4\delta k \frac{G'}{G} \frac{\gamma'^2 - 1}{\gamma'^2} A_x - \delta \frac{\varrho'}{\varrho} \frac{k^2 - \varkappa_t^2}{k} A_x = 0,$$

$$H_2(a, b) + \delta H_1(a, b) - i \frac{\varrho'}{\varrho} \delta \frac{k^2 - \varkappa_t^2}{k} A_z = 0$$

with

$$H_1(a, b) = (k^2 + \varkappa_t^2) a + 2k\varkappa_l b, \qquad H_2(a, b) = 2k\varkappa_t a + (k^2 + \varkappa_t^2) b.$$

Making allowance for (14) and (20) this yields a linear system for a and b:

$$(\alpha_{11} + \delta\beta_{11}) a + (\alpha_{12} + \delta\beta_{12}) b = 0, \qquad (\alpha_{21} + \delta\beta_{21}) a + (\alpha_{22} + \delta\beta_{22}) b = 0$$

where

$$\alpha_{11} = \alpha_{22} = k^2 + \varkappa_t^2, \qquad \alpha_{12} = 2k\varkappa_l, \qquad \alpha_{21} = 2k\varkappa_t,$$

$$\beta_{11} = -\frac{\gamma'^2 - 2}{\gamma'^2} 2k\varkappa_t + 4 \frac{G'}{G} \frac{\gamma'^2 - 1}{\gamma'^2} k\varkappa_t - \frac{\varrho'}{\varrho} \frac{k^2 - \varkappa_t^2}{k^2} k\varkappa_t,$$

$$\beta_{12} = -\frac{\gamma'^2 - 2}{\gamma'^2} (k^2 + \varkappa_t^2) + 4 \frac{G'}{G} \frac{\gamma'^2 - 1}{\gamma'^2} k^2 - \frac{\varrho'}{\varrho} (k^2 - \varkappa_t^2),$$

$$\beta_{21} = k^2 + \varkappa_t^2 - \frac{\varrho'}{\varrho} (k^2 - \varkappa_t^2), \qquad \beta_{22} = 2k\varkappa_l - \frac{\varrho'}{\varrho} \frac{k^2 - \varkappa_t^2}{k} \varkappa_l.$$

For solubility the determinant must vanish. In the first order we get

$$\Phi(\omega, k) \equiv (k^2 + \varkappa_t^2)^2 - 4k^2\varkappa_t\varkappa_l \equiv \alpha_{11}\alpha_{22} - \alpha_{12}\alpha_{21} = -\delta(\alpha_{11}\beta_{22} + \alpha_{22}\beta_{11} - \alpha_{12}\beta_{21} - \alpha_{21}\beta_{12}).$$

From this equation (27) immediately follows.

Appendix 2

Higher order expansion

In generalizing (17) we write for the displacements in the surface layer:

$$\vec{u}' = \sum_{n=0}^{\cdots} \vec{A}^{(n)} (kz)^n \, e^{i(kx-\omega t)} .$$

For the film stresses we get with (23)

$$\sigma'_{xx} = G'k\{\gamma'^2 \, i \sum_0 A_x^{(n)} (kz)^n + (\gamma'^2 - 2) \sum_1 n A_z^{(n)} (kz)^{n-1}\} \, e^{i(kx-\omega t)} ,$$

$$\sigma'_{xz} = G'k \, \{\sum_1 n A_x^{(n)} (kz)^{n-1} + i \sum_0 A_z^{(n)} (kz)^n\} \, e^{i(kx-\omega t)} ,$$

$$\sigma'_{zz} = G'k\{i(\gamma'^2 - 2) \sum_0 A_x^{(n)} (kz)^n + \gamma'^2 \sum_1 n A_z^{(n)} (kz)^{n-1}\} \, e^{i(kx-\omega t)} .$$

The boundary conditions at free surface $z = d$ yield (in analogy to (24)):

$$0 = \sum_1 n A_x^{(n)} \delta^{n-1} + i \sum_0 A_z^{(n)} \delta^n , \qquad 0 = \sum_0 i(\gamma'^2 - 2) \, A_x^{(n)} \delta^n + \gamma'^2 \sum_1 n A_z^{(n)} \delta^{n-1} .$$

The continuity conditions at boundary $z = 0$ remain unchanged. In the present notation (25) reads

$$H_1(a, b) = \frac{G'}{G} \, k(A_x^{(1)} + i A_z^{(0)})$$

$$H_2(a, b) = \frac{G'}{G} \, ik(i \, (\gamma'^2 - 2) \, A_x^{(0)} + \gamma'^2 A_z^{(1)}) .$$

In extension of (26) the averaged first moments of the equation of motion (22) and (34) are evaluated to

$$-\omega^2 d\varrho' \sum_0 A_x^{(n)} \frac{1}{n+l+1} \delta^{n+l} = -\delta_{0l} G'k(A_x^{(1)} + i A_z^{(0)}) +$$

$$+ G'k \left\{ -\gamma'^2 \sum_0 \frac{1}{n+l+1} A_x^{(n)} \delta^{n+l+1} + i(\gamma'^2 - 2) \sum_1 \frac{n}{n+l} A_z^{(n)} \delta^{n+l} \right\} -$$

$$- G'k \left\{ \sum_1 \frac{nl}{n+l-1} A_x^{(n)} \delta^{n+l-1} + i \sum_0 \frac{l}{n+l} A_z^{(n)} \delta^{n+l} \right\}$$

$$-\omega^2 d\varrho' \sum_0 A_z^{(n)} \frac{1}{n+l+1} \delta^{n+l} = -G'k(A_x^{(1)} + i A_z^{(0)}) \, \delta_{0l} +$$

$$+ G'k \left\{ i \sum_1 \frac{n}{n+l} A_x^{(n)} \delta^{n+l} - \sum_0 \frac{1}{n+l+1} A_z^{(n)} \delta^{n+l+1} \right\} -$$

$$- G'k \left\{ i(\gamma'^2 - 2) \sum_0 A_x^{(n)} \frac{l}{n+l} \delta^{n+l} + \gamma'^2 \sum_1 A_z^{(n)} \frac{nl}{n+l-1} \delta^{n+l-1} \right\} .$$

BOOK REVIEWS

Brams, S. J. / Lucas, W. F. / Straffin, P. D., jr. (eds.), Political and Related Models. Reprint. Berlin-Heidelberg-New York, Springer-Verlag 1983. XX, 396 S., 28 Abb., DM 72,—. US $ 29.80. ISBN 3-540-90696-7 (Modules in Applied Mathematics 2)

Lucas, F. / Roberts, F. S. / Thrall, R. M. (eds.), Discrete System Models. Reprint. Berlin-Heidelberg-New York, Springer-Verlag 1983. XX, 353 S., 133 Abb., DM 72,—, US $ 31.10. ISBN 3-540-90724 (Modules in Applied Mathematics 3)

Marcus-Roberts, H. / Thompson, M. (eds.), Life Science Models. Reprint. Berlin-Heidelberg-New York, Springer-Verlag 1983. XX, 366 S., 80 Abb., DM 72,—. US $ 28.60. ISBN 3-540-90739-4 (Modules in Applied Mathematics 4)

Dem Referenten liegen die Bände 2, 3 und 4 der 4-Band-Serie Modules in Applied Mathematics vor. Das Anliegen dieser Serie ist es, die Wirkungsweise der modernen Mathematik zur Lösung aktueller praktischer Probleme sichtbar zu machen. Dazu wurden 63 Einzelarbeiten verschiedener Autoren in vier Bänden thematisch zusammengefaßt. Band 2 beschäftigt sich in 14 Kapiteln mit Problemen aus Politik, Soziologie und Ökologie, Band 3 mit ebenfalls 14 Kapiteln betrachtet Modelle aus Graphentheorie, Kombinatorik und Netzwerkplanung, und in Band 4 werden Populationsmodelle, Probleme aus der Biomedizin und noch einmal Probleme aus der Ökologie behandelt. Der hier nicht referierte Band 1 befaßt sich mit gewöhnlichen und partiellen Differentialgleichungsmodellen.

Es sei gleich zu Beginn herausgestellt, daß es Herausgebern und Autoren sehr gut gelungen ist, die vorliegende Serie als einheitliches Ganzes vorzulegen, einheitlich nach inhaltlichem Anspruch und nach äußerer Form und Realisierung. Jedes Kapitel besteht aus Problembeschreibung, mathematischem Modell, Algorithmen, Analysis, detaillierten Beispielen, Bemerkungen über Implementation auf einem Computer, Bemerkungen zu Erweiterungen und Übungen. Dabei wird besonderer Wert auf die Modellierungsphase gelegt. Hier liegt auch die Schlüsselrolle des Gesamtwerkes, Mittler zwischen praktischem Problem und Kalkül zu sein.

Das umfangreiche Werk sei im folgenden inhaltlich durch die Angabe der Kapitelüberschriften und der Autoren charakterisiert.

Band 2: The Process of Applied Mathematics (M. Thompson), Proportional Representation (E. Bolger), Comparison Voting (S. J. Brams), Modeling Coalitional Values (W. F. Lucas, L. J. Billera), Urban Wastewater Management Planning (J. P. Heaney), An Everyday Approach to Matrix Operations (R. M. Thrall, E. L. Perry), Sources of Applications of Mathematics in Ecological and Environmental Subject Areas, Suitable for Classroom Use (H. T. Hayslett, Jr.), How To Ask Sensitive Questions without Getting Punched in the Nose (J. C. Maceli), Measuring Power in Wieghted Voting Systems (W. F. Lucas), To the (Minimal Winning) Victors Go the (Equally Divided) Spoils: A New Power Index for Simple n-Person Games (J. Deegan, Jr., E. W. Packel), Power Indices in Politics (P. D. Straffin, Jr.), Committee Decision Making (P. Rice), Stochastic Difference Equations with Sociological Applications (L. Cobb), The Apportionment Problem (W. F. Lucas).

Band 3: Foresight — Insight — Hindsight (J. C. Frauenthal, T. L. Saaty), Five Nautical Models (R. L. Baker), An Optimal Inventory Policy Model (O. Marrero), An Arithmetic Model of Gravity (D. Greenspan), Four-Way Stop of Traffic Lights? An Illustration of the Modeling Process (E. W. Packel), A Model for Municipal Street Sweeping Operations (A. C. Tucker, L. Bodin), Finite Covering Problems (R. E. Prather), A Pulse Process Model of Athletic Financing (E. L. Perry), Traffic Equilibria on a Roadway Network (E. L. Peterson), An Optimal Mix Problem: Capacity Expansion in a Power Field (J. Zahavi), Hierarchies, Reciprocal Matrices, and Ratio Scales (T. L. Saaty), Multiple-Choice Testing (R. J. Weber), Computing Fixed Points, with Applications to Economic Equilibrium Models (M. J. Todd), Production Costing and Reliability Assessment of Electrical Power Generation Systems under Supply and Demand Uncertainty (J. Zahavi).

Band 4: Population Mathematics (F. C. Hoppensteadt), Population Growth: An Age Structure Model (R. E. Fennell), A Comparison of Some Deterministic and Stochastic Models of Population Growth (H. Marcus-Roberts), Some Examples of Mathematical Models for the Dynamics of Several-Species Ecosystems (H. R. van der Vaart), Malaria: Models of the Population Dynamics of the Malaria Parasite (H. Marcus-Roberts, F. S. Roberts), MacDonald's Work on Helminth Infections (D. Ludwig, B. D. Haystock), A Model for the Spread of Gonorrhea (M. Braun), DNA, RNA, and Random Mating: Simple Applications of the Multiplication Rule (H. Marcus-Roberts), Cigarette Filtration (D. A. Drew), Efficiency of Energy Use in Obtaining Food, I: Humans (F. S. Roberts), Efficiency of Energy Use in Obtaining Food, II: Animals (F. S. Roberts, H. Marcus-Roberts), The Spatial Distribution of Cabbage Butterfly Eggs (D. L. Solomon). Alle Bände der Serie sind bestens geeignet für Vorlesungen, Seminare und Selbststudium in allgemeinen mathematischen Grundkursen und besonders in Kursen zur mathematischen Modellierung auf unterschiedlichem Niveau. Dabei kann ein Einsatz als tragende und/oder Ergänzungsliteratur erfolgen. Umfangreiche Literaturangaben erleichtern und fördern weitergehende Untersuchungen und Einordnungen.

Dresden J. Terno

Fletcher, C. A. J., Computational Galerkin Methods. Berlin-Heidelberg-New York-Tokyo, Springer-Verlag 1984. XI, 309 S., DM 88,—. US £ 32.90. ISBN 3-540-12633-3. (Springer Series in Computational Physics)

In dieser Monographie ist es dem Autor gelungen, die vielfältigen numerischen Methoden unter einem einheitlichen Gesichtspunkt zu ordnen und übersichtlich darzustellen. Dabei spielen das Galerkin-Verfahren und seine Verallgemeinerungen eine zentrale Rolle. Neben dem klassischen Galerkin-Verfahren und den Spektralmethoden werden Methoden der finiten Elemente, Randintegralmethoden, Differenzenverfahren, Kollokationsmethoden, Methoden der kleinsten Quadrate, Momentenmethoden und Pseudospektralmethoden dargestellt und einige an konkreten Beispielen miteinander verglichen. Bei der Lösung spezieller Anwendungsaufgaben werden unterschiedliche Koordinaten- und Testfunktionen eingeführt, Kombinationen unterschiedlicher Methoden betrachtet und ihr Einfluß auf die Genauigkeit, Konvergenz, Stabilität und Effektivität der Rechenverfahren untersucht. Ausgehend von einem historischen Überblick werden zunächst einfache Beispiele und dann mit der Weiterentwicklung der Methoden auch kompliziertere Aufgaben gelöst. Durch die klare Gliederung des Buches werden die Vorteile der einzelnen Methoden ihre Leistungsfähigkeit und ihre Grenzen deutlich herausgearbeitet und Wege zur effektiven Lösung zahlreicher Anwendungsaufgaben dargestellt. Im einzelnen werden gewöhnliche und partielle Differentialgleichungen der Hydro-, Aero-, Thermo- und Elektrodynamik behandelt und konkrete Aufgaben untersucht, wie z. B. die Strömung im Kanal, Wärmeleitung im Stab, freie Konvektion im Rechteck, hydrodynamische Stabilität, Schallausbreitung in Rohren, Umströmung eines Tragflügels, kompressible Umströmung eines Kreiszylinders, elektrische Ladungsverteilung in einer Kreisscheibe, konvektiver Wärmetransport, Strahlklappenströmung, Gezeitenströmung, Wettervorhersage, Temperatur- und Druckverteilungen in der Atmosphäre und turbulente Strömungen. Dabei werden die Approximationen der linearen und nichtlinearen Differentialoperatoren ausführlich betrachtet, während die rechentechnischen Probleme bei der Lösung großer algebraischer Gleichungssysteme und funktionalanalytische Untersuchungen der Modellgleichungen nicht Hauptanliegen dieser Monographie sind.

Jedes der 7 Kapitel des Buches enthält ein Literaturverzeichnis und einen abschließenden Abschnitt, in dem über das jeweilige Kapitel und die Verbindung zu anderen Kapiteln referiert und auf weitere Details der Methoden und Anwendungsaufgaben hingewiesen wird. Außer dem Sachregister werden am Ende des Buches 2 FORTRAN-Programme zur Lösung der Burgers-Gleichung publiziert und ihre Benutzung erläutert.

Diese Monographie ist insbesondere Ingenieuren und angewandten Mathematikern sowie Studenten zu empfehlen, die sich einen Überblick über die neueste Entwicklung der modernen numerischen Methoden und die rechentechnische Lösung konkreter Probleme verschaffen möchten oder selbst konkrete Aufgaben aus der Technik oder den Naturwissenschaften zu lösen haben.

Berlin W. Zwick

ZAMM · Z. Angew. Math. u. Mech. 65 (1985) 1, 29–36

PANAGIOTOPOULOS, P. D.

Nonconvex Problems of Semipermeable Media and Related Topics[1])

In dieser Arbeit werden Randwertprobleme behandelt, die im Rahmen der Theorie der semipermeablen Medien für nicht konvexe Permeabilitätspotentiale formuliert werden. Semipermeabilitätsbedingungen werden im Inneren und auf dem Rand des Körpers berücksichtigt. Nach der Formulierung der Hemivariationsungleichungen der Probleme und der Definition der entsprechenden regularisierten Probleme wird die Existenz der Lösung bewiesen. Der Beweis ist konstruktiv und stellt eine Methode zur Approximation der Lösung dar. Abschließend wird die entwickelte Methode auf ähnliche Probleme der Elastizitätstheorie angewandt.

The present paper deals with BVPs in the theory of semipermeability potentials. Both the cases of interior and boundary semipermeability problems are studied. After the formulation of the problems in terms of hemivariational inequalities and the definition of the regularised problems an existence result is proved. The proof is constructive and may be used for the approximation of the solution of the problem. Finally, the developed method is applied to certain similar problems arising in the theory of elasticity.

В настоящей статье рассмотрены краевые задачи из теории полупроницаемых средств, сформулированные для невыпуклых потенциалов проницаемости. Полупроницаемости учтены внутри тела и на его крае. После формулировки проблем в виде хеми-вариационных неравенств и определения соответствующих регуляризованных проблем доказано существование решения. Доказательностьво конструктивно и может использоваться как метод аппроксимации решения. Наконец, развитый здесь метод применяется к аналогичным задачам из теории упругости.

1. Introduction

Semipermeability problems were first studied in [1] in the context of convexity, i.e. for monotone semipermeability relations. These relations lead to variational inequalities connected with the Δ-operator and arise in heat conduction in electrostatics and in flow problems through porous media. They concern the behaviour of semipermeable membranes either on the boundary or in the interior of the body under consideration.

To analogous variational inequalities lead certain categories of temperature control problems in heat conduction, or potential control and pressure control in electrostatics and hydraulics, respectively, which concern monotone regulation of the respective quantity on the boundary or in the interior of the body. Here, we shall study the same problems but without assuming monotonicity. The arising B.V.Ps lead to hemivariational inequalities and substationarity problems [2] [3], instead of variational inequalities and minimum problems as happens in the convex case, since the potentials are nonconvex. After the formulation of the B.V.P. the corresponding hemivariational inequalities are studied concerning the existence of the solution and a method for the approximation of the solution is proposed.

The developed theory does not hold for the Δ-operator only but also for other elliptic operators. In this context certain applications from the theory of elasticity are also discussed.

2. Formulation of the problem

We consider further an open, bounded, connected subset Ω of $\mathbb{R}^3$ referred to a fixed Cartesian coordinate system $Ox_1x_2x_3$ and we formulate the equation

$$-\Delta u = f \quad \text{in } \Omega \tag{2.1}$$

for stationary problems and for instationary problems the equation (normalized diffusion equation)

$$\frac{\partial u}{\partial t} - \Delta u = f \quad \text{in } \Omega\,. \tag{2.2}$$

Here t denotes the time and u represents the temperature in the case of heat conduction problems, whereas in problems of hydraulics and electrostatics the pressure and the electric potential are represented, respectively. We denote further by Γ the boundary of Ω and we assume that Γ is sufficiently regular ($C^{1,1}$-boundary is sufficient). If $n = \{n_i\}$ denotes the outward unit normal to Γ then $\partial u/\partial n$ is the flux of heat, fluid or electricity through Γ for the aforementioned classes of problems, respectively. Concerning the derivation of (2.1) (2.2) and the underlying assumptions, see [1]. Further, we shall follow the language of heat conduction problems.

We may consider two main classes of *semipermeability problems*: the *interior* and the *boundary semipermeability problems* [1].

In the first class of problem the classical boundary conditions

$$u = 0 \quad \text{on} \quad \Gamma \tag{2.3}$$

[1]) This paper contains results of the author's research during Summersemester 1981, when he was as a Visiting Professor at the Institute of Applied Mathematics of the University of Hamburg.

are assumed to hold whereas in the second class the boundary conditions are defined as a relation between $\partial u/\partial n$ and u. In the first class the semipermeability conditions are obtained by assuming that $f = \bar{f} + \bar{\bar{f}}$ where $\bar{\bar{f}}$ is given and $\bar{f}$ is a known function of u. Here, we consider (2.3) for the sake of simplicity, but the proposed method holds for every other classical boundary condition. G. Duvaut and J. L. Lions [1] have considered cases of monotone semipermeability problems. All these problems may be put in the following general framework. For the first class we seek a function u such as to satisfy (2.1) with

$$f = \bar{f} + \bar{\bar{f}}, \qquad -\bar{f} \in \partial j_1(u) \quad \text{in } \Omega \tag{2.4}$$

and (2.3). In the instationary case (2.1) is replaced by (2.2) and additionally the initial condition

$$u|_{t=0} = u_0 \tag{2.5}$$

must hold. Here $j_1: R \to (-\infty, +\infty]$, $j_1 \not\equiv \infty$, is a convex, lower semicontinuous functional on R and ∂ denotes the symbol of subdifferentiation [4]. It is well-known (see e.g. [5]) that $\partial j_1(u) = \beta_1(u)$ where $\beta_1: R \to R$ denotes a maximal monotone operator (i.e. $(\xi, \beta_1(\xi))$ is the graph of a complete nondecreasing curve in R^2). For the second class we seek a function u such that (2.1) is satisfied together with the boundary condition

$$-\frac{\partial u}{\partial n} \in \partial j_2(u) \quad \text{on } \Gamma \tag{2.6}$$

where j_2 has the same properties as j_1. The dynamic problem is defined by (2.2), (2.5) and (2.6). Note, that if $q = \{q_i\}$ denotes the heat flux vector and $k > 0$ is the coefficient of thermal conductivity of the material we may write by Fourier's law that $q_i n_i = -k\, \partial u/\partial n$.

To give an example of (2.4) we consider the diagram of fig. 1a. It corresponds to the condition

$$-\bar{f} \in \partial I_K(u) \tag{2.4a}$$

where $K = \{u/u \geqq h\}$ and $I_K(u) = \{0 \text{ if } u \in K,\ \infty \text{ if } u \notin K\}$ (the indicator of K) or equivalently to the condition

$$u \geqq h, \qquad \bar{f} \geqq 0, \qquad (u - h)\,\bar{f} = 0 \tag{2.4b}$$

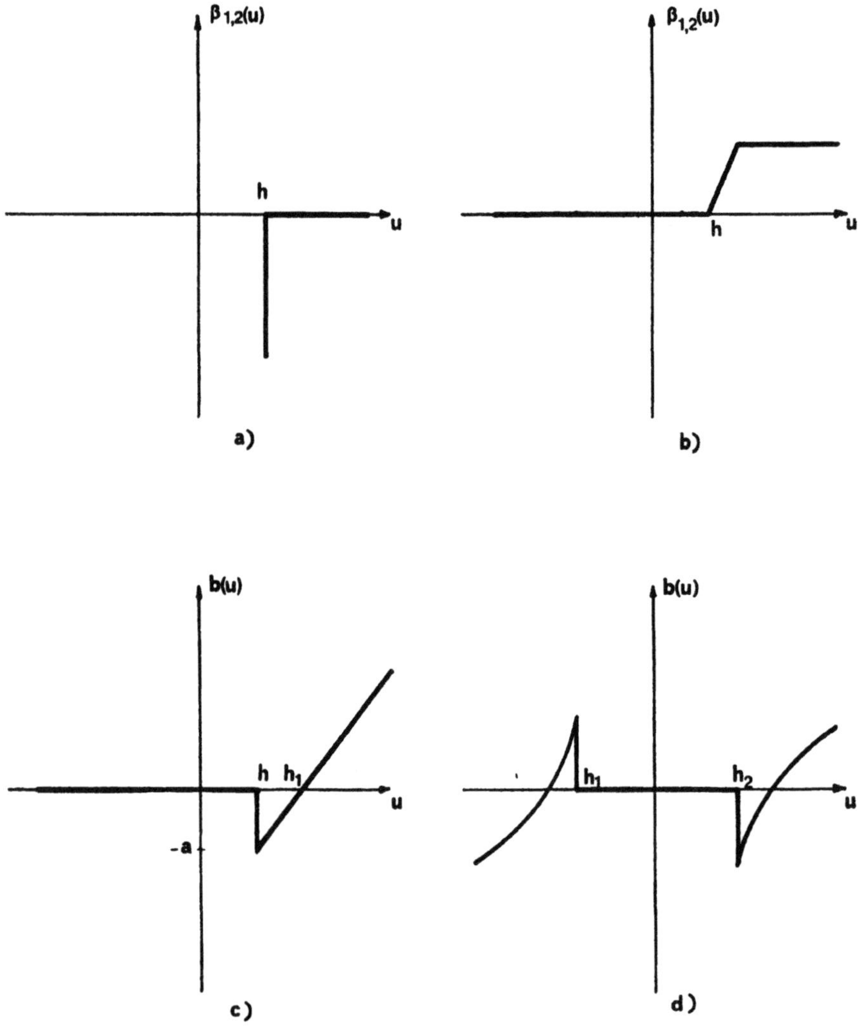

Fig. 1. Semipermeability and control relations

which says that if the temperature is equal to h then an additional heating $\bar{f}$ takes place through a servomechanism. The same graph, with β_1 replaced by β_2, describes the behaviour of a semipermeable membrane of negligible thickness [1]. The graph of fig. 1b describes a control problem in which the temperature is regulated by thermostatic control devices to deviate as little as possible from the interval $(-\infty, +h]$. A natural generalization is to consider analogous relations to the preceding ones but nonmonotone. For instance, the relation

$$-\frac{\partial u}{\partial n} = b(u) \quad \text{on } \Gamma, \tag{2.7}$$

where $b: R \to R$ is given in fig. 1c and corresponds to the behaviour of a semipermeable membrane (e.g. a wall) of finite thickness:

If $u < h$ the heat flux tends to leave Ω but the semipermeable membrane does not permit it, i.e. $\partial u/\partial n = 0$. If $u \geqq h$ a heat flux enters into Ω through Γ and we may assume that $-\partial u/\partial n = \alpha(u - h_1)$, $\alpha > 0$, $h_1 > h$. Analogous relations may be considered in the case of internal semipermeability. The graph of fig. 1d corresponds to a temperature control problem in which we regulate the temperature to deviate as little as possible from the interval $[h_1, h_2]$.

We, thus, are led to consider more general conditions of the form

$$f = \bar{f} + \bar{\bar{f}}, \; -\bar{f} \in b_1(u) \quad \text{in } \Omega \tag{2.8}$$

and

$$\frac{1}{k} q_i n_i = -\frac{\partial u}{\partial n} \in b_2(u) \quad \text{on } \Gamma \tag{2.9}$$

instead of (2.4) and (2.6), respectively. Here b_1 and b_2 are general multivalued functions from R into R (i.e. they include jumps) which are nonmonotone. Now we state the following B.V.Ps. (classical formulation).

Problem 1: Find u such as to satisfy (2.1) (or (2.2) and (2.5) in the instationary case), (2.8) and the boundary condition (2.3).

Problem 2: Find u such as to satisfy (2.1) (or (2.2) and (2.5) in the instationary case) and the boundary condition (2.9).

3. Variational formulations

We assume that b (b_1 or b_2): $R \to R$ is a locally bounded measurable function in R, i.e. $b \in L^\infty_{\text{loc}}(R)$. Then for any $\varepsilon > 0$ and $\xi \in R$ we define the numbers

$$\bar{b}_\varepsilon(\xi) = \operatorname*{ess\,sup}_{|\xi_1 - \xi| < \varepsilon} b(\xi) \quad \text{and} \quad \underline{b}_\varepsilon(\xi) = \operatorname*{ess\,inf}_{|\xi_1 - \xi| < \varepsilon} b(\xi) \tag{3.1}$$

which are decreasing and increasing functions of ε, respectively. Thus we may formulate the lim $\bar{b}_\varepsilon(\xi) = \bar{b}(\xi)$ and lim $\underline{b}_\varepsilon(\xi) = b(\xi)$ as $\varepsilon \to 0$. For instance, in the case of the jump discontinuity of fig. 1c $\bar{b}(h) = 0$ and $\underline{b}(h) = -a$. Thus we define, following [6], the multivalued function $\hat{b}: R \to R$ defined by setting

$$\hat{b}(\xi) = [\underline{b}(\xi), \bar{b}(\xi)]\,. \tag{3.2}$$

It is shown in [7] that if $b(\xi_{\pm 0})$ exists for every $\xi \in R$ then a locally Lipschitz function $\varphi: R \to R$ exists such that

$$\hat{b}(\xi) = \bar{\partial}\varphi(\xi) \tag{3.3}$$

where φ is defined by means of the relation

$$\varphi^0(\xi, z) = \limsup_{\substack{h \to 0 \\ \lambda \to 0+}} \frac{1}{\lambda} \int_{\xi+h}^{\xi+h+\lambda z} b(\xi)\, d\xi\,. \tag{3.4}$$

Here $\bar{\partial}$ denotes the generalized gradient of CLARKE [8] and $\varphi^0(\xi, z)$ the directional derivative in the sense of CLARKE at ξ in the direction z [9].

Thus (2.8) and (2.9) take the form

$$f = \bar{f} + \bar{\bar{f}}, \qquad -\bar{f} \in \bar{\partial}\varphi_1(u) \quad \text{in } \Omega \tag{3.5}$$

and

$$\frac{1}{k} q_i n_i = -\frac{\partial u}{\partial n} \in \bar{\partial}\varphi_2(u) \quad \text{on } \Gamma\,. \tag{3.6}$$

Boundary conditions and material laws connected with the $\bar{\partial}$-operator (the so-called *superpotential* in the sense of CLARKE) have been introduced in [2]. Note that by definition (3.5) and (3.6) are equivalent to the "hemivariational" inequalities [2] [3]

$$\varphi_1^0(u, v - u) \geqq -\bar{f}(v - u) \quad \text{for all} \quad v \in R\,, \qquad f = \bar{f} + \bar{\bar{f}} \tag{3.7}$$

and

$$\varphi_2^0(u, v - u) \geqq \frac{1}{k} q_i n_i(v - u) \quad \text{for all} \quad v \in R . \tag{3.8}$$

From (2.1) we obtain by multiplying by $v - u$, integrating on Ω and applying (formally) the Green-Gauss theorem (we set $\int\limits_\Omega uv \, d\Omega = (u, v)$ and $a(u, v) = \int\limits_\Omega \operatorname{grad} u \cdot \operatorname{grad} v \, d\Omega$)

$$\left(\frac{\partial u}{\partial t}, v - u\right) + a(u, v - u) = \int\limits_\Gamma \frac{\partial u}{\partial n}(v - u) \, d\Gamma + (f, v - u) . \tag{3.9}$$

The combination of (3.9) with (3.7) (resp. (3.8)) leads to the following (formal) variational problems.

Problem 1′: Find u such as to satisfy (2.3) and the hemivariational inequality

$$a(u, v - u) + \int\limits_\Omega \varphi_1^0(u, v - u) \, d\Omega \geqq (\bar{\bar{f}}, v - u) \tag{3.10}$$

for every v with $v = 0$ on Γ.

Problem 2′: Find u such as to satisfy the hemivariational inequality

$$a(u, v - u) + \int\limits_\Gamma \varphi_2^0(u, v - u) \, d\Gamma \geqq (f, v - u) \tag{3.11}$$

for every v.

Note that if φ_1 and φ_2 are convex, then $\bar{\partial}\varphi_1 = \partial\varphi_1$ and $\partial\varphi_2 = \partial\varphi_2$ and therefore (3.10) and (3.11) reduce to the well-known variational inequalities of [1].

For the respective instationary cases u satisfies for the interior semipermeability problem, (2.3), (2.5) and the inequality

$$\left(\frac{\partial u}{\partial t}, v - u\right) + a(u, v - u) + \int\limits_\Omega \varphi_1^0(u, v - u) \, d\Omega \geqq (\bar{\bar{f}}, v - u) \tag{3.12}$$

for every v with $v = 0$ on Γ, and for the boundary semipermeability problem, (2.5) and the inequality

$$\left(\frac{\partial u}{\partial t}, v - u\right) + a(u, v - u) + \int\limits_\Gamma \varphi_2^0(u, v - u) \, d\Gamma \geqq (f, v - u) \tag{3.13}$$

for every v.

For the formulation of the stationary problems as substationarity problems we refer to [10].

4. Functional setting and existence results

We shall prove two propositions on the existence of the solution of problems 1 and 2.

For problem 1 (resp. 2) we assume that $u \in \mathring{H}^1(\Omega)$ (resp. $u \in H^1(\Omega)$), $\bar{\bar{f}} \in H^{-1}(\Omega)$ (resp. $f \in (H^1(\Omega))'$). Now we take a function $\varrho \in C_0^\infty(-1, 1)$ with $\varrho \geqq 0$ and $\int\limits_{-\infty}^{+\infty} \varrho(\xi) \, d\xi = 1$ (mollifier) and we consider the new function $\varrho_\varepsilon(\xi) = \frac{1}{\varepsilon} \varrho\left(\frac{\xi}{\varepsilon}\right)$ and the convolution

$$b_\varepsilon = \varrho_\varepsilon * b , \qquad \varepsilon > 0 , \tag{4.1}$$

$b_\varepsilon \in C^\infty(R)$ is a regularized form of b. We assume also that

$$\operatorname*{ess\,sup}_{(-\infty, -\xi)} b \leqq \operatorname*{ess\,inf}_{(\xi, +\infty)} b . \tag{4.2}$$

Without loss of generality we may assume[2]) that

$$0 \in (\operatorname*{ess\,sup}_{(-\infty, -\xi)} b, \operatorname*{ess\,inf}_{(\xi, +\infty)} b) . \tag{4.2a}$$

Moreover, let $Au = -\frac{\partial}{\partial x_i}\left(a_{ij}\frac{\partial u}{\partial x_j}\right)$, $i, j = 1, \dots, n$, $a_{ij} = a_{ji} \in L^\infty(\Omega)$ and $a_{ij}(x) \, \xi_i \xi_j \geqq c|\xi|^2$, $c > 0$, for all $\xi \in R^n$ for a.e. $x \in R^n$ (e.g. $A = -\Delta$) and let

$$a(u, v) = \int\limits_\Omega a_{ij} \frac{\partial u}{\partial x_i} \frac{\partial v}{\partial x_j} \, d\Omega .$$

---

[2]) Indeed we may determine c such that $c \in (b(-\xi), b(+\xi))$ for a.e. ξ with $|\xi| > \xi_0$. Through the replacement of b by $b - c$ and of $\bar{\bar{f}}$ by $\bar{\bar{f}} - c$ we may assume that (4.2a) holds.

We consider now the following problems:

Problem 1'': For any $\bar{\bar{f}} \in H^{-1}(\Omega)$ find $u \in \mathring{H}^1(\Omega)$ and an $\bar{f} \in H^{-1}(\Omega) \cap L^1(\Omega)$ such that

$$Au = \bar{f} + \bar{\bar{f}} \quad \text{a.e. in } \Omega \tag{4.3}$$

and

$$-\bar{f}(x) \in \hat{b}_1(u(x)) = \bar{\partial}\varphi_1(u(x)) \quad \text{a.e. in } \Omega . \tag{4.4}$$

Problem 1''': For any $\bar{\bar{f}} \in H^{-1}(\Omega)$ find $u \in \mathring{H}^1(\Omega)$ such that

$$a(u, v - u) + \int_\Omega \varphi_1^0(u, v - u) \, d\Omega \geqq (\bar{\bar{f}}, v - u) \quad \text{for all} \quad v \in \mathring{H}^1(\Omega) . \tag{4.5}$$

Problem $1_\varepsilon''$: For any $\bar{\bar{f}} \in H^{-1}(\Omega)$ find $u_\varepsilon \in \mathring{H}^1(\Omega)$ and $\bar{f} \in H^{-1}(\Omega) \cap L^1(\Omega)$ such that

$$Au_\varepsilon = \bar{f} + \bar{\bar{f}} \quad \text{a.e. in } \Omega \tag{4.6}$$

and

$$-\bar{f}(x) = b_{1\varepsilon}(u_\varepsilon(x)) = \frac{d\varphi_{1\varepsilon}}{du}(u_\varepsilon(x)) \quad \text{a.e. in } \Omega . \tag{4.7}$$

Problem $1_\varepsilon'''$: For any $\bar{\bar{f}} \in H^{-1}(\Omega)$ find $u_\varepsilon \in \mathring{H}^1(\Omega)$ such that

$$a(u, v) + \left(\frac{d\varphi_{1\varepsilon}}{du}(u_\varepsilon(x)), v\right) = (\bar{\bar{f}}, v) \quad \text{for all} \quad v \in \mathring{H}^1(\Omega) . \tag{4.8}$$

Problems $1_\varepsilon''$ and $1_\varepsilon'''$ are equivalent and are respectively the regularized form of the equivalent Problems 1'' and 1''' which correspond to the classically formulated Problems 1 and 1'. The following proposition holds (cf. also [6]).

Proposition 1: *Suppose that* $b_1 \in L^\infty_{\text{loc}}(R)$ *and that* (4.2) *holds. Then there exists a solution to problem* 1''', *which is obtained from a solution* u_ε *of problems* $1_\varepsilon''$ *and* $1_\varepsilon'''$ *as* $\varepsilon \to 0$.

Proof: First we shall prove by means of a Galerkin approximation the existence of a solution u_ε of problem $1_\varepsilon'''$. Let us consider a Galerkin basis $\{w_i\}$ of $V = \mathring{H}^1(\Omega) \cap L^\infty(\Omega)$ and let V_n be a corresponding n-dimensional subspace of V. We consider the finite dimensional problem: find $u_{\varepsilon n} \in V_n$ such that

$$a(u_{\varepsilon n}, v) + (b_{1\varepsilon}(u_{\varepsilon n}), v) - (\bar{\bar{f}}, v) = 0 \quad \text{for all} \quad v \in V_n . \tag{4.9}$$

This problem has a solution $u_{\varepsilon n}$ as we may show by applying Brouwer's fixed point theorem. In particular we apply the following lemma [12]. If $\tilde{\Lambda}: R^n \to R^n$ is continuous such that for $r > 0$ $(\tilde{\Lambda}a)_i a_i \geqq 0$ for all $a \in R^n$ with $|a| = r$, then there exists a_0 with $|a_0| \leqq r$ such that $\tilde{\Lambda}a_0 = 0$.

We write now (4.9) in the form

$$(\Lambda u_{\varepsilon n}, v) = 0 \quad \text{for all} \quad v \in V_n \tag{4.10}$$

where $\Lambda: V_n \to V_n'$. If $u_{\varepsilon n} - \sum_{i=1}^n a_i w_i$ corresponds to $a = \{a_i\}$ we pose

$$(\tilde{\Lambda}a)_i = a(u_{\varepsilon n}, w_i) + (b_{1\varepsilon}(u_{\varepsilon n}), w_i) - (\bar{\bar{f}}, w_i) . \tag{4.11}$$

We may determine $\varrho_1 > 0$ and $\varrho_2 > 0$ such that $b_{1\varepsilon}(\xi) \geqq 0$ if $\xi > \varrho_1$, $b_{1\varepsilon}(\xi) \leqq 0$ if $\xi < -\varrho_1$ and $|b_{1\varepsilon}(\xi)| \leqq \varrho_2$, if $|\xi| \leqq \varrho_1$. Then (4.11) implies that

$$(\Lambda u_{\varepsilon n}, u_{\varepsilon n}) \geqq c_1 \|u_{\varepsilon n}\|^2_{\mathring{H}_1} + \int_{|u_{\varepsilon n}| > \varrho_1} u_{\varepsilon n} b_{1\varepsilon}(u_{\varepsilon n}) \, d\Omega + \int_{|u_{\varepsilon n}| \leqq \varrho_1} u_{\varepsilon n} b_{1\varepsilon}(u_{\varepsilon n}) \, d\Omega +$$

$$- \|\bar{\bar{f}}\|_{H^{-1}} \|u_{\varepsilon n}\|_{\mathring{H}^1} \geqq c_1 \|u_{\varepsilon n}\|^2_{\mathring{H}^1} - c_2 \|u_{\varepsilon n}\|_{\mathring{H}^1} + c_3 . \tag{4.12}$$

Thus an $r > 0$ can be determined such that

$$(\Lambda u_{\varepsilon n}, u_{\varepsilon n}) \geqq 0 \quad \text{for} \quad \|u_{\varepsilon n}\|_{\mathring{H}^1} = r \tag{4.13}$$

and thus (4.9) has a solution with $\|u_{\varepsilon n}\|_{H^1} \leqq r$.

Since $\{u_{\varepsilon n}\}$ is bounded in $\mathring{H}^1(\Omega)$ we may consider a subsequence, again denoted by $\{u_{\varepsilon n}\}$, such that as $\varepsilon \to 0$ and $n \to \infty$

$$u_{\varepsilon n} \to u \quad \text{weakly in } \mathring{H}^1(\Omega) \tag{4.14}$$

and

$$u_{\varepsilon n} \to u \quad \text{strongly in } L^2(\Omega) \tag{4.15}$$

by the Rellich-Kondrachov theorem. Thus

$$u_{\varepsilon n} \to u \quad \text{a.e. in } \Omega . \tag{4.16}$$

Further we shall show that $\{b_{1\varepsilon}(u_{\varepsilon n})\}$ is weakly precompact in $L^1(\Omega)$. We have thus to show that for each $\alpha > 0$ there exists a δ such that if $\text{mes}\,\omega < \delta$ then $\int_\omega |b_{1\varepsilon}(u_{\varepsilon n})| \, d\Omega < \alpha$. To this end we apply to $b_{1\varepsilon}$ the obvious inequality [6]

$$\xi_0 |p(\xi)| \leqq |\xi p(\xi)| + \xi_0 \sup_{|\xi| \leqq \xi_0} |p(\xi)| . \tag{4.17}$$

We find

$$\int_\omega |b_{1\varepsilon}(u_{\varepsilon n})| \, d\Omega \leqq \frac{1}{\xi_0} \int_\omega |b_{1\varepsilon}(u_{\varepsilon n}) \, u_{\varepsilon n}| \, d\Omega + \int_\omega \sup_{|u_{\varepsilon n}| \leqq \xi_0} |b_{1\varepsilon}(u_{\varepsilon n})| \, d\Omega . \tag{4.18}$$

From (4.9) and the boundedness of $\{u_{\varepsilon n}\}$ we deduce easily that $\int_\omega |b_{1\varepsilon}(u_{\varepsilon n}) \, u_{\varepsilon n}| \, d\Omega$ is bounded. We choose in (4.18) ξ_0 large enough, so that the first term in the right-hand side of (4.18) is to be $< \alpha/2$ for all n, and let δ be such that

$$\operatorname{ess\,sup}_{|\xi| < \xi_0 + 1} |b_{1\varepsilon}(\xi)| \leqq \frac{\alpha}{2\delta} . \tag{4.19}$$

Thus we obtain the weak precompactness of $\{b_{1\varepsilon}(u_{\varepsilon n})\}$ in $L^1(\Omega)$. Accordingly a subsequence can be determined such that as $\varepsilon \to 0$ and $n \to \infty$

$$b_{1\varepsilon}(u_{\varepsilon n}) \to \chi \quad \text{weakly in } L^1(\Omega) . \tag{4.20}$$

By taking the limit in (4.9) we obtain easily, since $v \in L^\infty(\Omega)$ and $\bigcup V_n$ is dense in $\mathring{H}^1(\Omega)$ the equality

$$\alpha(u, v) + (\chi, v) - (\bar{\bar{f}}, v) = 0 \quad \text{for all} \quad v \in \mathring{H}^1(\Omega) \tag{4.21}$$

where

$$\chi \in H^{-1}(\Omega) \cap L^1(\Omega) .$$

To complete the proof we have to show that

$$\chi \in \hat{b}_1(u) = \bar{\partial}\varphi_1(u) \quad \text{a.e. in } \Omega \tag{4.22}$$

or because of (3.2) that

$$\chi \in [\underline{b}_1(u(x)), \bar{b}_1(u(x))] \quad \text{a.e. in } \Omega . \tag{4.22a}$$

For $\alpha > 0$ we may choose $\omega \subset \Omega$ with mes $\omega < \alpha$ such that due to (4.16)

$$u_{\varepsilon n} \to u \quad \text{uniformly in } \Omega - \omega \tag{4.23}$$

where u is essentially bounded on $\Omega - \omega$. Thus we may determine for any $\delta > 0$ an n_0 and an ε_0 such that for $n > n_0$ and $\varepsilon < \varepsilon_0$

$$|u_{\varepsilon n}(x) - u(x)| < \frac{\delta}{2} \quad \text{for all} \quad x \in \Omega - \omega ; \tag{4.24}$$

$b_{1\varepsilon}(u_{\varepsilon n}(x))$ is an average function of b_1 for $n > n_0$ and $\varepsilon < \varepsilon_0$ and for

$$-\frac{\delta}{2} + u_{\varepsilon n}(x) < \xi < \frac{\delta}{2} + u_{\varepsilon n}(x) , \qquad x \in \Omega - \omega .$$

Since this range of ξ is a subset of $[-\delta + u_{\varepsilon n}(x), \delta + u_{\varepsilon n}(x)]$ we obtain that

$$b_{1\varepsilon}(u_{\varepsilon n}(x)) \in \hat{b}_1(u(x)) = \bar{\partial}\varphi_1(u(x)) . \tag{4.25}$$

Similarly we obtain for $e \geqq 0$ a.e. in $\Omega - \omega$ and essentially bounded, that

$$\int_{\Omega - \omega} \underline{b}_{1\delta}(u_{\varepsilon n}(x)) \, e \, d\Omega \leqq \int_{\Omega - \omega} b_{1\varepsilon}(u_{\varepsilon n}(x)) \, e \, d\Omega \leqq \int_{\Omega - \omega} \bar{b}_{1\delta}(u_{\varepsilon n}(x)) \, e \, d\Omega \tag{4.26}$$

which implies in the limit that

$$\int_{\Omega - \omega} \underline{b}_{1\delta}(u(x)) \, e \, d\Omega \leqq \int_{\Omega - \omega} \chi \, e \, d\Omega \leqq \int_{\Omega - \omega} \bar{b}_{1\delta}(u(x)) \, e \, d\Omega \tag{4.27}$$

and since u is bounded on $\Omega - \omega$, (4.27) holds also as $\delta \to 0$ with $\underline{b}_{1\delta}$ and $\bar{b}_{1\delta}$ replaced by $\underline{b}_1$ and $\bar{b}_1$, respectively. It results that

$$\chi \in \bar{b}_1(u(x)) = \bar{\partial}\varphi_1(u(x)) \quad \text{a.e. in } \Omega - \omega \tag{4.29}$$

and by taking α as small as possible we obtain (4.22), q.e.d.

Analogously to the Problems 1″ and 1‴ and the regularized problems $1_\varepsilon''$ and $1_\varepsilon'''$ we may formulate the problems 2″, 2‴, $2_\varepsilon''$ and 2‴. In this case, due to the semicoercivity of $a(u, u)$ the existence proof is still open. Note that if $a(u, u)$ is strictly coercive, i.e. if

$$a(u, u) + c_1 |u|^2_{L^2} \geqq c_2 \|u\|^2_{H^1} , \qquad c_1 > 0 , \qquad c_2 > 0 \quad \text{for all} \quad u \in H^1(\Omega) \tag{4.30}$$

then we can show as before the existence of a solution $u \in H^1(\Omega)$ of the inequality

$$a(u, v - u) + c_1(u, v - u) + \int_\Gamma \varphi_2^0(u, v - u) \, d\Gamma \geqq (f, v - u) \quad \text{for all} \quad v \in H^1(\Omega) . \tag{4.30}$$

Proposition 2: *Suppose that* $b_2 \in L^\infty_{\text{loc}}(R)$ *and that* (4.2) *holds. Then for any* $f \in H^1(\Omega)$ *there exists a* $u \in H^1(\Omega)$ *such that* (4.30) *holds.*

The proof is a slight modification of the proof of prop. 1 and is here ommited.

5. Related topics in linear elasticity

Nonmonotone relations of the form depicted in figs. 1 c, d may be formulated also in the framework of other technical problems. Let us consider a Kirchhoff plate occupying in its undeformed state an open, bounded, connected subset Ω of R^2 which is referred to a fixed right-handed Cartesian coordinate system $0x_1x_2x_3$. The middle plane of the plate, assumed to have a constant thickness h, coincides with the $0x_2x_3$-plane. Let Γ be the boundary of Ω which is assumed to be appropriately regular. We denote by ζ the deflection of the points $x \in \Omega$. The plate is subjected to a distributed load $f = (0, 0, f_3)$, $f_3 = f_3(x)$ per unit area of the middle surface. We assume here that the boundary conditions

$$\zeta = \frac{\partial \zeta}{\partial n} = 0 \quad \text{on } \Gamma \tag{5.1}$$

hold. More complicated boundary conditions may be also treated. We assume, furthermore, that

$$f_3 = \bar{f}_3 + \bar{\bar{f}}_3 \tag{5.2}$$

where $\bar{\bar{f}}_3 \in L^2(\Omega)$ is a given load distribution and that on $\Omega' \subset \Omega$ with $\bar{\Omega}' \cap \Gamma = \emptyset$ the condition

$$-\bar{f}_3 \in \hat{b}(\zeta) = \bar{\partial}\varphi(\zeta) \tag{5.3}$$

holds, where $\hat{b}$ is defined by (3.2) and φ is given by (3.4). On $\Omega - \Omega'$ we assume that $\bar{f}_3 = 0$. Relation (5.3) describes a unilateral contact of the plate with a granular support which causes the nonmonotone behaviour (fig. 2a), or a brittle fracture effect or a delamination effect (fig. 2b).

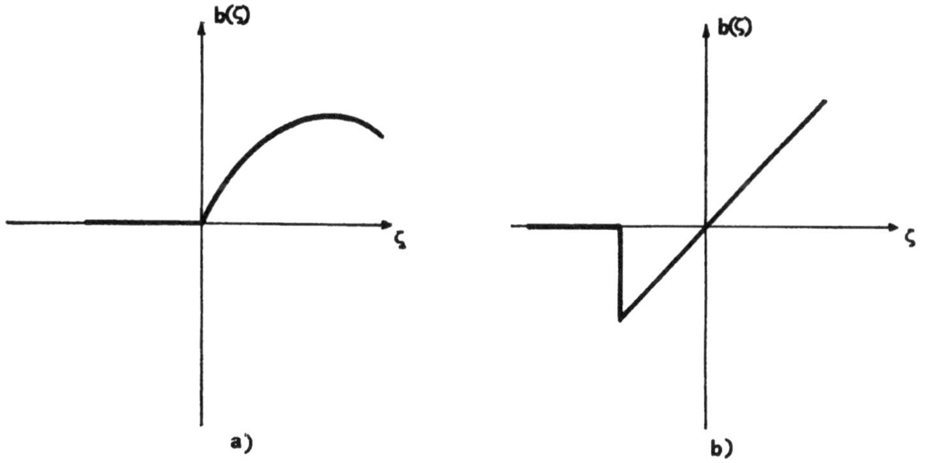

Fig. 2. Nonmonotone reaction – deflection laws in plate theory

Assuming that $\zeta \in \overset{\circ}{H}{}^2(\Omega)$ we obtain by applying the Green-Gauss theorem the expression (see e.g. [1] [10])

$$a(\zeta, z) = (f_3, z) \quad \text{for all} \quad z \in \overset{\circ}{H}{}^2(\Omega) \tag{5.4}$$

where $\frac{1}{2} a(\zeta, \zeta)$ denotes the elastic energy of the plate and $(f_3, z) = \int_\Omega f_3 z \, d\Omega$. We have

$$a(\zeta, z) = \frac{Eh^3}{12(1-\nu^2)} \int_\Omega \left((1-\nu) \frac{\partial^2 \zeta}{\partial x_\alpha \, \partial x_\beta} \frac{\partial^2 z}{\partial x_\alpha \, \partial x_\beta} + \nu \Delta\zeta \, \Delta z \right) d\Omega \,, \qquad 1 \leqq \alpha \,, \qquad \beta \leqq 2 \,, \qquad 0 < \nu < \frac{1}{2} \,, \tag{5.5}$$

where E is the modulus of elasticity and ν is the Poisson ratio. (5.4) combined with (5.2) and (5.3) leads to the following problem.

Problem 3: For any $\bar{\bar{f}}_3 \in L^2(\Omega)$ find $\zeta \in \overset{\circ}{H}{}^2(\Omega)$ such as to satisfy the hemivariational inequality

$$a(\zeta, z - \zeta) + \int_{\Omega'} \varphi^0(\zeta, z - \zeta) \, d\Omega' \geqq (\bar{\bar{f}}_3, z - \zeta) \quad \text{for all} \quad z \in \overset{\circ}{H}{}^2(\Omega) \,, \tag{5.6}$$

As in sec. 4 we may formulate the regularized problem:

Problem 3_ε: For any $\bar{\bar{f}}_3 \in L^2(\Omega)$ find $\zeta_\varepsilon \in \overset{\circ}{H}{}^2(\Omega)$ such as to satisfy the variational equality

$$a(\zeta, z) + \int_{\Omega'} \frac{d\varphi_\varepsilon(\zeta_\varepsilon)}{d\zeta} z \, d\Omega' = (\bar{\bar{f}}_3, z) \quad \text{for all} \quad z \in \overset{\circ}{H}{}^2(\Omega) \,. \tag{5.7}$$

The operator form of (5.7) is obvious. If φ is convex, (5.6) reduces to a variational equality [1]. The following proposition holds.

Proposition 3: *Suppose that $b \in L^\infty_{\mathrm{loc}}(\mathbb{R})$ and that* (4.2) *holds. Then Problem* 3 *admits a solution $\zeta \in \mathring{H}^2(\Omega)$ which is obtained from the solution of Problem* 3_ε *as* $\varepsilon \to 0$.

Proof: Considering the coercivity of $a(\zeta, \zeta)$ we can proceed[3]) as in the proof of prop. 1. Conditions analogous to (4.14), (4.15), (4.16), (4.20) are obtained, q.e.d.

Analogous B.V.Ps may be formulated in the framework of the theory of bars and the theory of von Kármán plates.

References

1 Duvaut, G.; Lions, J. L., Les inéquations en mécanique et en physique, Dunod, Paris 1972.

2 Panagiotopoulos, P. D., Nonconvex superpotentials in the sense of F. H. Clarke and applications, Mech. Res. Comm. 8 (1981), 335–340.

3 Panagiotopoulos, P. D., Nonconvex energy functionals. Application to nonconvex elastoplasticity, Mech. Res. Comm. 9 (1982), 23–29.

4 Rockafellar, R. T., Convex Analysis, Princeton University Press, Princeton 1970.

5 Panagiotopoulos, P. D., Ungleichungsprobleme in der Mechanik, RWTH Aachen 1977.

6 Rauch, J., Discontinuous semilinear differential equations and multiple valued maps, Proc. Amer. Math. Soc. 64 (1977), 277–282.

7 Chang, K. C., Variational methods for nondifferentiable functionals and their applications to partial differential equations, J. Math. Anal. Appl. 80 (1981), 102–129.

8 Clarke, F. H., Generalized gradients and applications, Trans. Amer. Math. Soc. 205 (1975), 247–262.

9 Rockafellar, R. T., Generalized directional derivatives and subgradients of nonconvex functions, Canadian J. Math. 32 (1980), 257–280.

10 Panagiotopoulos, P. D., Inequality problems in mechanics and applications. The convex and the nonconvex case. Birkhäuser Verlag, Basel Boston, 1983

11 Jeggle, H., Nichtlineare Funktionalanalysis, B.G. Teubner, Stuttgart 1979.

12 Lions, J. L., Quelques méthodes de résolution des problèmes aux limites nonlinéaires, Dunod-Gauthier Villars, Paris 1969.

Address: P. D. Panagiotopoulos, ord. Professor, School of Technology, Aristotle University Thessaloniki, Greece and Honorarprofessor, Faculty of Mathematics and Natural Sciences, RWTH Aachen, D-5100 Aachen, F.R.G.

---

[3]) Note that $\mathring{H}^2(\Omega) \subset L^\infty(\Omega)$ ([12] p. 45) for $\Omega \subset \mathbb{R}^2$

ZAMM · Z. Angew. Math. u. Mech. 65 (1985) 1, 37–47

STREIT, U.; WEINELT, W.

Fehlerabschätzung für eine Differenzenmethode zur Approximation schwacher Lösungen von Stefan-Aufgaben

In der Arbeit wird eine implizite Differenzenmethode (Gitterparameter h, τ) zur Berechnung der schwachen Lösung einer eindimensionalen Stefan-Aufgabe betrachtet, wobei die Enthalpiefunktion geeignet approximiert wird (Parameter Δ). Unter schwachen Regularitätsvoraussetzungen werden Fehlerabschätzungen für diese Methode gefunden. Hauptresultat ist eine Konvergenzordnung von $O(\Delta^{1/4} + h^{1/2} + \tau^{1/2})$ in der L_2-Norm.

The paper deals with an implicit finite difference method (parameters h, τ) for calculating the weak solution of a one-dimensional Stefan problem, where the enthalpy function is approximated in a suitable manner (parameter Δ). Error estimates are obtained under weak regularity assumptions. The main result is an L_2-convergence order of $O(\Delta^{1/4} + h^{1/2} + \tau^{1/2})$.

В работе рассматривается разностный метод с опережением (параметры h, τ) для аппроксимации слабого решения одномерной задачи Стефана. При этом функция энтальпии аппроксимируется подходящим образом (параметр Δ). При слабых условиях относительно регулярности решения получены оценки погрешности этого метода. Главный результат — порядок сходимости $O(\Delta^{1/4} + h^{1/2} + \tau^{1/2})$ в норме пространства L_2.

1. Aufgabenstellung; Einleitung

Im Gebiet $\Omega_T = \Omega \times \{t: 0 < t \leqq T\}$, $\Omega = \{x: 0 < x < 1\}$ wird folgende Randwertaufgabe für eine nichtlineare Wärmeleitungsgleichung betrachtet:

$$\left.\begin{aligned}
&\frac{\partial H}{\partial t} = \frac{\partial^2 u}{\partial x^2} + f(u)\,, && (x,t) \in \Omega_T\,,\\
&-\frac{\partial u}{\partial x} + \alpha_1\big(u - g_1(t)\big) = 0\,, && x = 0\,, \quad 0 < t \leqq T\,,\\
&\frac{\partial u}{\partial x} + \alpha_2\big(u - g_2(t)\big) = 0\,, && x = 1\,, \quad 0 < t \leqq T\,,\\
&H = H_0(x)\,, && x \in \bar{\Omega}\,, \quad t = 0\,.
\end{aligned}\right\} \tag{1.1}$$

Der Zusammenhang zwischen der Temperatur $u(x, t)$ und der Enthalpie $H(x, t)$ wird durch eine „ideale" Materialgleichung in Gestalt der mehrdeutigen Funktion

$$\left.\begin{aligned}
&H(x,t) = a\big(u(x,t)\big)\,, \qquad a(\xi) = \begin{cases} c_1\xi\,, & \xi < 0\\ c_2\xi + \lambda\,, & \xi > 0\end{cases}\\
&a(-0) \leqq a(0) \leqq a(+0)\,, \qquad c_i > 0\,, \quad i = 1, 2\,, \quad \lambda > 0
\end{aligned}\right\} \tag{1.2}$$

vermittelt. Durch (1.1) und (1.2) wird die Wärmeleitung in einem „reinen" Medium der konstanten (auf Eins normierten) Dichte und Wärmeleitfähigkeit beschrieben, das für negative (positive) Temperaturen als Phase 1 (Phase 2) mit der spezifischen Wärmekapazität $c_1(c_2)$ vorliegt. Bei $u = 0$ vollzieht sich eine Phasenumwandlung, wobei die spezifische latente Wärme λ umgesetzt wird.

Für die Funktionen H_0, g_1, g_2 und f, die die Anfangsenthalpieverteilung, die Umgebungstemperaturen und innere Wärmequellen repräsentieren, gelte vorerst

$$H_0 \in L_\infty(\Omega)\,, \qquad g_i \in L_\infty(0, T)\,, \qquad i = 1, 2\,,$$
$$|f(\xi') - f(\xi'')| \leqq L|\xi' - \xi''| \quad \text{für} \quad \xi', \xi'' \in R^1;$$

α_1 und α_2 sind die nichtnegativen Wärmeübergangskoeffizienten.

Es sei vermerkt, daß die durch $u(x, t) = a^{-1}\big(H(x, t)\big)$ festgelegte Umkehrfunktion

$$a^{-1}(\xi) = \begin{cases} c_1^{-1}\xi\,, & \xi < 0\\ 0\,, & 0 \leqq \xi < \lambda\\ c_2^{-1}(\xi - \lambda)\,, & \lambda \leqq \xi \end{cases}$$

von $a(.)$ existiert, eindeutig ist und im weiteren eine wichtige Rolle spielt.

Unter einer *schwachen Lösung* der Aufgabe (1.1), (1.2) wird ein Paar von Funktionen $u \in L_\infty(\Omega_T)$, $H \in L_\infty(\Omega_T)$ verstanden, das über (1.2) voneinander abhängt und das für beliebige Testfunktionen

$$\Phi \in C^{2,1}(\bar{\Omega}_T) \quad \text{mit} \quad \begin{cases} \Phi = 0 & \text{für} \quad t = T\,,\\ -\dfrac{\partial \Phi}{\partial x} + \alpha_1 \Phi = 0 & \text{für} \quad x = 0\,,\\ \dfrac{\partial \Phi}{\partial x} + \alpha_2 \Phi = 0 & \text{für} \quad x = 1 \end{cases}$$

der Identität

$$\iint_{\Omega_T} \left(H \frac{\partial \Phi}{\partial t} + u \frac{\partial^2 \Phi}{\partial x^2} + f(u)\, \Phi \right) \mathrm{d}x\, \mathrm{d}t + \int_0^1 H_0 \Phi(x, 0)\, \mathrm{d}x + \int_0^T \alpha_1 g_1 \Phi(0, t)\, \mathrm{d}t + \int_0^T \alpha_2 g_2 \Phi(1, t)\, \mathrm{d}t = 0 \quad (1.3)$$

genügt.

Die schwache Lösung ist eindeutig (siehe z. B. [1]). In der Literatur werden schwache Lösungen für Aufgaben des Typs (1.2), (1.3) etwa seit 1960 intensiver betrachtet. Neben der Untersuchung ihrer qualitativen Eigenschaften (Eindeutigkeit, Existenz, Regularität und Zusammenhang mit klassischen Lösungen von Stefan-Aufgaben, monotone Abhängigkeit, asymptotisches Verhalten) spielte dabei von Anfang an ihre Approximation durch numerische Methoden eine wichtige Rolle. Bis heute ist eine Vielzahl von Verfahren entwickelt worden, die sowohl auf Formulierungen in u als auch auf solchen in H basieren, die sowohl von expliziten als auch von impliziten Zeitdiskretisierungen ausgehen und die sich sowohl des Konzeptes der Differenzenmethoden als auch desjenigen der Finite-Element-Methoden bedienen. Stellvertretend seien hier [2], [3], [4], [5] genannt. Auch Zugänge über Variationsungleichungen sind bekannt (z. B. [6]). Schwierigkeiten, die sich wegen der Unstetigkeit von $a(.)$ speziell bei impliziten Varianten für die numerische Behandlung der algebraischen Aufgaben (nichtlineare Gleichungs- bzw. Ungleichungssysteme) ergeben, können umgangen werden, wenn $a(.)$ durch eine stetige oder sogar glatte Funktion $a_\Delta(.)$ approximiert wird ([7], [8]; dem entspricht eine „Verschmierung" der Phasenumwandlung über ein endliches Temperaturintervall $|u| < \Delta$, was in praktischen Anwendungen sogar eine gegenüber (1.2) realistischere Beschreibung der Materialgleichung darstellen kann). Auch hierfür sind verschiedene Differenzen- und Finite-Element-Methoden publiziert worden (z. B. [9], [10]; für eine detailliertere Übersicht und eine gewisse Wertung vorhandener Methoden siehe etwa [11] oder [12]).

In der Mehrzahl der Fälle wird die Konvergenz der Methoden bewiesen. Jedoch sind eigentlich keine Aussagen über deren Ordnung bzw. Fehlerabschätzungen bekannt geworden: Einige wenige Arbeiten geben eine Konvergenzordnung gegen die Lösung der mit Hilfe von $a_\Delta(.)$ geglätteten Aufgabe an, wobei zudem von hohen Regularitätsvoraussetzungen über diese Lösung ausgegangen wird, die nicht als theoretisch abgesichert gelten können (z. B. [9]). Ordnungsaussagen zur Konvergenz in Abhängigkeit vom Parameter Δ fehlen ganz und gar.

Diese Tatsachen kennzeichnen generell die gegenwärtige Situation auf diesem wichtigen Teilgebiet der Untersuchung numerischer Methoden für freie Randwertaufgaben der Wärmeleitungsgleichung (für einen Überblick hierzu siehe [13]).

Das Ziel der vorliegenden Arbeit besteht darin, für eine implizite Differenzenmethode zur Approximation einer schwachen Lösung von Aufgabe (1.2), (1.3) eine Fehlerabschätzung anzugeben, die auf bekannten Regularitätseigenschaften der Lösung basiert. Dazu werden in Abschnitt 2 Resultate über die schwache Lösung und ihre Approximation aus [1] zitiert. In Abschnitt 3 wird der Einfluß des Parameters Δ auf das Konvergenzverhalten untersucht, während die Konvergenz der Differenzenmethode in Abhängigkeit von den Gitterparametern h und τ in Abschnitt 4 behandelt wird.

2. Existenz und Eigenschaften der schwachen Lösung und ihrer Approximationen

Für $\Delta > 0$ wird

$$a_\Delta(\xi) = \begin{cases} a(\xi)\,, & |\xi| > \Delta \\ \dfrac{a(\Delta) - a(-\Delta)}{2\Delta}\, \xi + \dfrac{a(\Delta) + a(-\Delta)}{2}\,, & |\xi| \leqq \Delta \end{cases} \tag{2.1}$$

definiert und über den Gittern

$$\left.\begin{aligned} &\omega_m = \omega \times \vartheta_m\,, && \omega = \{x_i = ih;\, i = 1, \dots, N-1;\, Nh = 1\}\,, \\ & && \vartheta_m = \{t_n = n\tau;\, n = 1, \dots, m \leqq M;\, M\tau = T\}\,, \\ &\overline{\omega}_m = \overline{\omega} \times \vartheta_m\,, && \overline{\omega} = \omega \cup \gamma\,, \quad \gamma = \{0, 1\} \end{aligned}\right\} \tag{2.2}$$

das Differenzenschema

$$\left.\begin{aligned} &(a_\Delta(y))_{\bar t} = y_{\bar x x} + f(y)\,, && (x, t) \in \omega_M\,, \\ &-y_x + \alpha_1(y - P_{\bar t} g_1(t)) = -\frac{h}{2}\left((a_\Delta(y))_{\bar t} - f(y)\right), && x = 0\,, \quad t \in \vartheta_M\,, \\ &y_{\bar x} + \alpha_2(y - P_{\bar t} g_2(t)) = -\frac{h}{2}\left((a_\Delta(y))_{\bar t} - f(y)\right), && x = 1\,, \quad t \in \vartheta_M\,, \\ &a_\Delta(y) = P_x^2 H_0(x)\,, && x \in \overline{\omega}\,, \quad t = 0 \end{aligned}\right\} \tag{2.3}$$

betrachtet. Dabei wurden allgemein übliche Bezeichnungen (vgl. z. B. [14], [15]) für Differenzenquotienten verwendet. Im weiteren werden Gitterfunktionen y auf der „Zeitschicht n" auch kurz mit y^n notiert. Mit P sind Mittelungsfunktionale

$$\left.\begin{aligned} &P_{\bar t}w(t) = \int_{-\infty}^{+\infty} \Phi_-(t'\,;t)\,w(t')\,\mathrm{d}t'\,, \qquad P_x^2 w(x) = \int_{-\infty}^{+\infty} \Phi_2(x'\,;x)\,w(x')\,\mathrm{d}x'\,,\\ &\Phi_-(t'\,;t) = \frac{1}{\tau}\begin{cases}1\,, & t' \in (t-\tau, t)\\ 0\,, & t' \notin (t-\tau, t)\,,\end{cases}\\ &\Phi_2(x'\,;x) = \frac{1}{h^2}\begin{cases}x'-x+h\,, & x' \in (x-h, x)\\ x-x'+h\,, & x' \in (x, x+h)\\ 0\,, & x' \notin (x-h, x+h)\end{cases} \end{aligned}\right\} \tag{2.4}$$

eingeführt.

Raumsymbole und Normen werden analog zum kontinuierlichen Fall verwendet (siehe [16], [17]). Ohne Index werden Skalarprodukt und Norm im Raum $L_2(\bar\omega)$ bezeichnet:

$$(u, v) = \sum_{i=1}^{N-1} hu(x_i)\,v(x_i) + \frac{h}{2}\left(u(x_0)\,v(x_0) + u(x_N)\,v(x_N)\right), \qquad \|u\| = (u, u)^{1/2}\,.$$

Geringfügig davon abweichend, werden gelegentlich auch benötigt

$$[u, v) = \sum_{i=0}^{N-1} hu(x_i)\,v(x_i),\ \|[u\| = [u, u)^{1/2}\,, \qquad (u, v] = \sum_{i=1}^{N} hu(x_i)\,v(x_i)\,, \qquad \|u]\| = (u, u]^{1/2}\,,$$

und für $u, v \in L_2(\bar\omega_m)$ bedeutet

$$(u, v)_{2,\bar\omega_m} = \sum_{n=1}^{m} \tau(u^n, v^n)\,, \qquad \|u\|_{2,\bar\omega_m} = (u, u)_{2,\bar\omega_m}^{1/2}\,.$$

Im folgenden werden einige Lösungseigenschaften für die Aufgaben (1.2), (1.3) und (2.1), (2.3) angegeben, die in [1] für eine allgemeinere Aufgabenklasse begründet wurden. Die Angaben erfolgen dabei für beliebige $0 < t \leqq T$ bzw. beliebige $0 < m \leqq M$.

1° *Für $c_0 - \tau L > 0$, $c_0 = \min\{c_1, c_2\}$, existiert eine eindeutige Lösung $y \in L_\infty(\bar\omega_m)$ von Aufgabe* (2.1), (2.3), *die mit dem nichtlinearen* SOR-*Verfahren* ([18]) *berechnet werden kann, wobei für Relaxationsparameter aus dem Intervall* (0, 1] *die Iterationen global konvergieren. Es ist*

$$\|y\|_{\infty,\bar\omega_m} \leqq \max\{\|a_\Delta^{-1}(P_x^2 H_0)\|_{\infty,\bar\omega}, \|P_{\bar t} g_1\|_{\infty,\vartheta_m}, \|P_{\bar t} g_2\|_{\infty,\vartheta_m}\} + \begin{cases}\dfrac{|f(0)|}{L}\exp\dfrac{Lt_m}{c_0-\tau L}\,, & L > 0\\ |f|\,t_m c_0^{-1}\,, & L = 0\,.\end{cases}$$

2° *Für $H_0 \in W_2^1(\Omega)$, $g_i \in W_1^1(0, T)$, $i = 1, 2$, gilt $y \in W_2^1(\bar\omega_m)$ mit*

$$\begin{aligned} c_0\|y_{\bar t}\|_{2,\bar\omega_m}^2 \leqq{}& c_0^{-2}\|[(P_x^2 H_0)_x\|^2 +\\ &+ 2\|y\|_{\infty,\bar\omega_m}\{\alpha_1(2\|P_{\bar t} g_1\|_{\infty,\vartheta_m} + \|(P_{\bar t} g_1)_{\bar t}\|_{1,\vartheta_m}) + \alpha_2(2\|P_{\bar t} g_2\|_{\infty,\vartheta_m} + \|(P_{\bar t} g_2)_{\bar t}\|_{1,\vartheta_m})\} +\\ &+ (\alpha_1+\alpha_2)\,\|y\|_{\infty,\bar\omega_m}^2 + c_0^{-1}\|f(y)\|_{2,\bar\omega_m}^2 + O(h)\,,\\ \|[y_x^m\|^2 \leqq{}& c_0^{-2}\|[(P_x^2 H_0)_x\|^2 +\\ &+ 2\|y\|_{\infty,\bar\omega_m}\{\alpha_1(2\|P_{\bar t} g_1\|_{\infty,\vartheta_m} + \|(P_{\bar t} g_1)_{\bar t}\|_{1,\vartheta_m}) + \alpha_2(2\|P_{\bar t} g_2\|_{\infty,\vartheta_m} + \|(P_{\bar t} g_2)_{\bar t}\|_{1,\vartheta_m})\} +\\ &+ (\alpha_1+\alpha_2)\,\|y\|_{\infty,\bar\omega_m}^2 + \frac{1}{2c_0}\|f(y)\|_{2,\bar\omega_m}^2 + O(h)\,. \end{aligned}$$

3° *Für $H_0 \in W_2^1(\Omega)$, $g_i \in W_1^1(0, T)$, $i = 1, 2$, existiert eine schwache Lösung $u \in W_2^{1,1}(\Omega_T)$, $H \in L_2(\Omega_T)$ von Aufgabe* (1.2), (1.3). *Darüberhinaus ist $u \in C(\Omega_T)$.*

Für die stückweise konstanten L_2-Fortsetzungen $\tilde y$ der Lösung y von Aufgabe (2.1), (2.3) *gilt bei $h, \tau, \Delta \to 0$:*

$$\|\tilde y - u\|_{2,\Omega_T} \to 0\,,$$

$$a_\Delta(\tilde y) \xrightarrow{L_2(\Omega_T)} H\,, \qquad \tilde y_t \xrightarrow{L_2(\Omega_T)} \frac{\partial u}{\partial t}\,, \qquad \tilde y_x \xrightarrow{L_2(\Omega_T)} \frac{\partial u}{\partial x}\,.$$

Aus 1°, 2° folgt für $h, \tau, \Delta \to 0$ unmittelbar

4° *Für die Lösung der Aufgabe* (1.2), (1.3) *gelten die Abschätzungen*

$$\left.\begin{aligned} &\|u\|_{\infty,\Omega_t} \leqq \max\{\|a^{-1}(H_0)\|_{\infty,\Omega}, \|g_1\|_{\infty,(0,t)}, \|g_2\|_{\infty,(0,t)}\} + \begin{cases}\dfrac{|f(0)|}{L}\exp(Lt/c_0)\,, & L > 0\\ |f|\,t/c_0\,, & L = 0\,,\end{cases}\\ &c_0\left\|\frac{\partial u}{\partial t}\right\|_{2,\Omega_t}^2 \leqq c_0^{-2}\|H_0'\|_{2,\Omega}^2 + 2\|u\|_{\infty,\Omega_t}\{\alpha_1(2\|g_1\|_{\infty,(0,t)} +\\ &\qquad + \|\dot g_1\|_{1,(0,t)}) + \alpha_2(2\|g_2\|_{\infty,(0,t)} + \|\dot g_2\|_{1,(0,t)})\} +\\ &\qquad + (\alpha_1+\alpha_2)\,\|u\|_{\infty,\Omega_t}^2 + c_0^{-1}\|f(u)\|_{2,\Omega_t}^2\,,\\ &\left\|\frac{\partial u}{\partial x}(\cdot, t)\right\|_{2,\Omega}^2 \leqq c_0^{-2}\|H_0'\|_{2,\Omega}^2 + 2\|u\|_{\infty,\Omega_t}\{\alpha_1(2\|g_1\|_{\infty,(0,t)} + \|\dot g_1\|_{1,(0,t)}) +\\ &\qquad + \alpha_2(2\|g_2\|_{\infty,(0,t)} + \|\dot g_2\|_{1,(0,t)})\} + (\alpha_1+\alpha_2)\,\|u\|_{\infty,\Omega_t}^2 + \frac{1}{2c_0}\|f(u)\|_{2,\Omega_t}^2\,. \end{aligned}\right\} \tag{2.5}$$

Der Fehler $\tilde{y} - u$ wird gemäß

$$\|\tilde{y} - u\|_{2,\Omega_T} \leqq \|u_\Delta - u\|_{2,\Omega_T} + \|\tilde{y} - u_\Delta\|_{2,\Omega_T}$$

zerlegt und (beide Anteile der rechten Seite für sich) in den Abschnitten 3 und 4 untersucht. Dabei ist u_Δ, H_Δ schwache Lösung von (1.3), (2.1). Alle Aussagen dieses Abschnittes über u, H können unmittelbar auf u_Δ, H_Δ übertragen werden; speziell ist $u_\Delta \in W_2^1(\Omega_T) \cap C(\Omega_T)$. Für festes $\Delta > 0$ gilt sogar $H_\Delta \in W_2^1(\Omega_T) \cap C(\Omega_T)$.

3. Konvergenzordnung in Abhängigkeit vom Parameter Δ

Es seien u, H und u_Δ, H_Δ schwache Lösungen von (1.2), (1.3) bzw. von (1.3), (2.1). Ihre Differenz genügt der Beziehung

$$\begin{aligned} 0 &= \iint_{\Omega_T} \left\{(H - H_\Delta)\frac{\partial \Phi}{\partial t} + (u - u_\Delta)\frac{\partial^2 \Phi}{\partial x^2} + \big(f(u) - f(u_\Delta)\big)\Phi\right\} \mathrm{d}x\,\mathrm{d}t = \\ &= \iint_{\Omega_T} (H - H_\Delta)\left\{\frac{\partial \Phi}{\partial t} + A(x,t)\frac{\partial^2 \Phi}{\partial x^2} + A(x,t)\,B(x,t)\,\Phi\right\} \mathrm{d}x\,\mathrm{d}t + \\ &\quad + \iint_{\Omega_T} \left(a^{-1}(H) - a_\Delta^{-1}(H)\right)\left(\frac{\partial^2 \Phi}{\partial x^2} + B(x,t)\,\Phi\right) \mathrm{d}x\,\mathrm{d}t \end{aligned} \tag{3.1}$$

mit

$$A(x,t) = \begin{cases} \dfrac{a_\Delta^{-1}(H) - a_\Delta^{-1}(H_\Delta)}{H - H_\Delta}, & H \neq H_\Delta \\ 0, & H = H_\Delta, \end{cases}$$

$$B(x,t) = \begin{cases} \dfrac{f(u) - f(u_\Delta)}{u - u_\Delta}, & u \neq u_\Delta \\ 0, & u = u_\Delta \end{cases}$$

für die oben bei (1.3) definierten Testfunktionen Φ.

Nach Voraussetzung ist $0 < \Delta \leqq A(x,t) \leqq c_0^{-1}$, $|B(x,t)| \leqq L$. Im weiteren wird gezeigt, daß

$$\iint_{\Omega_T} |H - H_\Delta|\,|V|\,\mathrm{d}x\,\mathrm{d}t = O(\Delta^{1/2})$$

gilt. Vorerst sei $V \in \mathring{C}_\infty(\Omega_T)$ mit

$$\|V\|_{\infty,\Omega_T} \leqq M_0\,, \qquad \left\|\frac{\partial V}{\partial x}(.,t)\right\|_{2,\Omega} \leqq M_1;$$

M_0, M_1-Konstanten (unabhängig von Δ).

Die Lösungen Φ_n der Randwertaufgabe

$$\left.\begin{array}{lll} \dfrac{\partial \Phi_n}{\partial t} + A_n\dfrac{\partial^2 \Phi_n}{\partial x^2} + A_nB_n\Phi_n = V\,, & 0 < x < 1\,, & 0 \leqq t < T\,, \\ \Phi_n = 0\,, & 0 \leqq x \leqq 1\,, & t = T\,, \\ -\dfrac{\partial \Phi_n}{\partial x} + \alpha_1\Phi_n = 0\,, & x = 0\,, & 0 \leqq t < T\,, \\ \dfrac{\partial \Phi_n}{\partial x} + \alpha_2\Phi_n = 0\,, & x = 1\,, & 0 \leqq t < T\,, \end{array}\right\} \tag{3.2}$$

wobei A_n, B_n glatte Funktionen mit

$$\Delta \leqq A_n \leqq c_0^{-1}\,, \qquad |B_n| \leqq L\,, \qquad \|A - A_n\|_{2,\Omega_T} \leqq \frac{1}{n}\,, \qquad \|AB - A_nB_n\|_{2,\Omega_T} \leqq \frac{1}{n}$$

sind, existieren in der Klasse der in Abschnitt 1 bei (1.3) definierten Testfunktionen und genügen den Ungleichungen

$$\|\Phi_n\|_{\infty,\Omega_T} \leqq M_2 \quad \text{mit} \quad M_2 = M_0T\exp\left(1 + LT/c_0\right) \tag{3.3}$$

(man sehe dazu die Sätze 2.2, 2.3 aus [17], Kapitel 1).

Multiplikation der Differentialgleichung in (3.2) mit der Funktion $\dfrac{\partial^2 \Phi_n}{\partial x^2}\,\mathrm{e}^{\Theta t}$, $\Theta > 0$, und Integration über Ω_T führt auf

$$\begin{aligned} &\iint_{\Omega_T} \mathrm{e}^{\Theta t}\left\{A_n\left(\frac{\partial^2 \Phi_n}{\partial x^2}\right)^2 + \Theta\left(\frac{\partial \Phi_n}{\partial x}\right)^2 + A_nB_n\Phi_n\frac{\partial^2 \Phi_n}{\partial x^2} + \frac{\partial V}{\partial x}\frac{\partial \Phi_n}{\partial x}\right\} \mathrm{d}x\,\mathrm{d}t = \\ &= -\frac{1}{2}\int_0^1 \left(\frac{\partial \Phi_n}{\partial x}(x,0)\right)^2 \mathrm{d}x - \frac{\Theta}{2}\int_0^T \mathrm{e}^{\Theta t}\left(\alpha_1\Phi_n^2(0,t) + \alpha_2\Phi_n^2(1,t)\right)\mathrm{d}t + \frac{1}{2}\left(\alpha_1\Phi_n^2(0,0) + \alpha_2\Phi_n^2(1,0)\right). \end{aligned}$$

Mit

$$\frac{\partial V}{\partial x}\frac{\partial \Phi_n}{\partial x} \leqq \frac{\varepsilon_1}{4}\left(\frac{\partial \Phi_n}{\partial x}\right)^2 + \frac{1}{\varepsilon_1}\left(\frac{\partial V}{\partial x}\right)^2, \qquad \varepsilon_1 > 0,$$

$$B_n\Phi_n \frac{\partial^2 \Phi_n}{\partial x^2} \leqq \frac{\varepsilon_2}{4}\left(\frac{\partial^2 \Phi_n}{\partial x^2}\right)^2 + \frac{1}{\varepsilon_2}(B_n\Phi_n)^2, \qquad \varepsilon_2 > 0,$$

folgt weiter

$$\iint_{\Omega_T}\left(1-\frac{\varepsilon_2}{4}\right)A_n\left(\frac{\partial^2\Phi_n}{\partial x^2}\right)^2 e^{\Theta t}\,dx\,dt + \iint_{\Omega_T}\left(\Theta - \frac{\varepsilon_1}{4}\right)\left(\frac{\partial \Phi_n}{\partial x}\right)^2 e^{\Theta t}\,dx\,dt \leqq$$

$$\leqq \frac{1}{\varepsilon_1}\iint_{\Omega_T}\left(\frac{\partial V}{\partial x}\right)^2 e^{\Theta t}\,dx\,dt + \frac{1}{\varepsilon_2}\iint_{\Omega_T} A_n(B_n\Phi_n)^2 e^{\Theta t}\,dx\,dt + \frac{1}{2}\left(\alpha_1\Phi_n^2(0,0) + \alpha_2\Phi_n^2(1,0)\right),$$

woraus man für $\Theta = \varepsilon_1/4 = 1/T$, $\varepsilon_2 = 1/2$ die Abschätzung

$$\left\|\sqrt{A_n}\frac{\partial^2\Phi_n}{\partial x^2}\right\|_{2,\Omega_T} \leqq M_3, \qquad M_3 = \left(4(M_1T)^2 + 2(LM_2)^2\,Tc_0^{-1} + (\alpha_1+\alpha_2)\,M_2\right)^{1/2} \tag{3.4}$$

erhält. In Gleichung (3.1) wird Φ durch Φ_n ersetzt und $\frac{\partial \Phi_n}{\partial t}$ mittels (3.2) ausgedrückt. Es entsteht

$$\iint_{\Omega_T}(H - H_\Delta)\,V\,dx\,dt = I_1 + I_2 + I_3 + I_4 .$$

Die Integrale

$$I_1 = \iint_{\Omega_T}\left(a_\Delta^{-1}(H) - a^{-1}(H)\right)\left(\frac{\partial^2\Phi_n}{\partial x^2}\right)^2 dx\,dt, \qquad I_2 = \iint_{\Omega_T}\left(a_\Delta^{-1}(H) - a^{-1}(H)\right)B\Phi_n\,dx\,dt,$$

$$I_3 = \iint_{\Omega_T}(H - H_\Delta)\,(A_n - A)\left(\frac{\partial^2\Phi_n}{\partial x^2}\right)^2 dx\,dt, \qquad I_4 = \iint_{\Omega_T}(H - H_\Delta)\,(A_nB_n - AB)\,\Phi_n\,dx\,dt$$

werden unter Beachtung von

$$|a_\Delta^{-1}(\xi) - a^{-1}(\xi)| \leqq \Delta \quad (\text{für } \xi \in R^1 \text{ beliebig}) \tag{3.5}$$

abgeschätzt:

$$|I_1| \leqq \iint_{\Omega_T}\frac{|a_\Delta^{-1}(H) - a^{-1}(H)|}{\sqrt{A_n}}\sqrt{A_n}\left|\frac{\partial^2\Phi_n}{\partial x^2}\right| dx\,dt \leqq \frac{\Delta}{\sqrt{\Delta}}\sqrt{T}\left\|\sqrt{A_n}\frac{\partial^2\Phi_n}{\partial x^2}\right\|_{2,\Omega_T} \leqq M_3T^{1/2}\Delta^{1/2},$$

$$|I_2| \leqq \Delta LM_2T,$$

$$|I_3| \leqq \|H - H_\Delta\|_{\infty,\Omega_T}\iint_{\Omega_T}\frac{A_n - A}{\sqrt{A_n}}\sqrt{A_n}\left|\frac{\partial^2\Phi_n}{\partial x^2}\right| dx\,dt \leqq$$

$$\leqq \left(\|H\|_{\infty,\Omega_T} + \|H_\Delta\|_{\infty,\Omega_T}\right)\left\|\frac{|A_n - A|}{\sqrt{A_n}}\right\|_{2,\Omega_T}\left\|\sqrt{A_n}\frac{\partial^2\Phi_n}{\partial x^2}\right\|_{2,\Omega_T} \leqq M_4\frac{1}{\sqrt{\Delta}\,n}M_3,$$

$$|I_4| \leqq \|H - H_\Delta\|_{\infty,\Omega_T}\,\|A_nB_n - AB\|_{2,\Omega_T}\sqrt{T}\,\|\Phi_n\|_{\infty,\Omega_T} \leqq M_4M_2\sqrt{T}\frac{1}{n}.$$

Für $n \to \infty$ werden $|I_3|$ und $|I_4|$ beliebig klein, so daß

$$\iint_{\Omega_T}|H - H_\Delta|\,|V|\,dx\,dt \leqq \Delta^{1/2}M_3T^{1/2} + \Delta LM_2T \tag{3.6}$$

gilt für beliebiges V mit $\|V\|_{\infty,\Omega_T} \leqq M_0$ und $\left\|\frac{\partial V}{\partial x}(.,t)\right\|_{2,\Omega} \leqq M_1$ für alle $0 < t \leqq T$.

Man wählt nun $V = u - u_\Delta$. Dann sind

$$M_0 = \|u\|_{\infty,\Omega_T} + \|u_\Delta\|_{\infty,\Omega_T}, \qquad M_1 = \left\|\frac{\partial u}{\partial x}(.,t)\right\|_{2,\Omega} + \left\|\frac{\partial u_\Delta}{\partial x}(.,t)\right\|_{2,\Omega} \tag{3.7}$$

tatsächlich von Δ unabhängig (siehe Abschnitt 2); und es gilt

$$\|u - u_\Delta\|_{2,\Omega_T}^2 = \iint_{\Omega_T}|a_\Delta^{-1}(H_\Delta) - a^{-1}(H_\Delta) + a^{-1}(H_\Delta) - a^{-1}(H)|\,|u_\Delta - u|\,dx\,dt \leqq$$

$$\leqq \iint_{\Omega_T}|a^{-1}(H_\Delta) - a^{-1}(H)|\,|u_\Delta - u|\,dx\,dt + \iint_{\Omega_T}|a_\Delta^{-1}(H_\Delta) - a^{-1}(H_\Delta)|\,|u_\Delta - u|\,dx\,dt \leqq$$

$$\leqq \frac{1}{c_0}\iint_{\Omega_T}|H - H_\Delta|\,|u_\Delta - u|\,dx\,dt + \Delta M_0T .$$

Damit ist unter Beachtung von (3.6) das folgende Resultat bewiesen.

Satz 1: *Unter den Voraussetzungen von 3°, Abschnitt 2, gilt die Abschätzung*

$$\|u - u_\Delta\|_{2,\Omega_T} \leqq \left(\frac{M_3 \sqrt{T}}{c_0}\right)^{1/2} \Delta^{1/4} + O(\Delta^{3/4})$$

mit den durch (3.3), (3.4), (3.7) *definierten Konstanten.*

Bemerkung: Die Konstanten M_i, $i = 0, \dots, 3$, sind gemäß (2.5) und den Bemerkungen am Ende von Abschnitt 2 nur von den Daten der Aufgabe (1.2), (1.3) abhängig, so daß durch Satz 1 eine a-priori-Fehlerabschätzung gegeben ist.

4. Konvergenzordnung in Abhängigkeit von den Gitterparametern h, τ

Nach Abschnitt 2 gelten für die Lösungen von (1.3), (2.1) die Aussagen $u_\Delta \in W_2^{1,1}(\Omega_T)$, $u_\Delta(.,t) \in W_2^1(0,1)$ für alle $0 < t \leqq T$ (und damit wegen der L-Stetigkeit von $H_\Delta = a_\Delta(u_\Delta)$ dieselben Eigenschaften auch bezüglich dieser Funktion). Daher kann als schwache Formulierung von (1.3), (2.1) die Integralidentität

$$\iint_{\Omega_T} \left(\frac{\partial H_\Delta}{\partial t} \Phi + \frac{\partial u_\Delta}{\partial x} \frac{\partial \Phi}{\partial x} - f(u_\Delta)\, \Phi\right) \mathrm{d}x\, \mathrm{d}t - \alpha_1 \int_0^T \big(g_1(t) - u_\Delta(0,t)\big)\, \Phi(0,t)\, \mathrm{d}t -$$

$$- \alpha_2 \int_0^T \big(g_2(t) - u_\Delta(1,t)\big)\, \Phi(1,t)\, \mathrm{d}t = 0 \quad \text{für alle} \quad \Phi \in W_2^{1,0}(\Omega_T); \qquad H_\Delta = a_\Delta(u_\Delta) \tag{4.1}$$

nebst der Anfangsbedingung

$$H_\Delta(x,0) = H_0(x) \tag{4.2}$$

dienen, die in offensichtlicher Weise aus der „schwächeren" Formulierung (1.3) hervorgeht.

Um eine Konvergenzordnung für die Differenzenmethode (2.3), (2.1) zur näherungsweisen Lösung der Aufgabe (1.3), (2.1) zu begründen, werden mit den bereits in (2.4) angegebenen speziellen Testfunktionen Φ und den entsprechenden Mittelungsfunktionalen die folgenden verallgemeinerten Bilanzgleichungen¹) gebildet:

$$\left.\begin{aligned}
&\big((P_x^2 a_\Delta(u_\Delta))_{\bar t} - (P_{\bar t} u_\Delta)_{\bar x x} - P_{\bar t} P_x^2 f(u_\Delta)\big)(x,t) = 0\,, && (x,t) \in \omega_M;\\
&\big((P_x^2 a_\Delta(u_\Delta))_{\bar t} - \frac{2}{h}(P_{\bar t} u_\Delta)_x + 2\alpha_1 h^{-1} P_{\bar t} u_\Delta - P_{\bar t} P_x^2 f(u_\Delta)\big)(0,t) = 2\alpha_1 h^{-1} P_{\bar t} g_1(t)\,, &&\\
&\big((P_x^2 a_\Delta(u_\Delta))_{\bar t} + \frac{2}{h}(P_{\bar t} u_\Delta)_{\bar x} + 2\alpha_2 h^{-1} P_{\bar t} u_\Delta - P_{\bar t} P_x^2 f(u_\Delta)\big)(1,t) = 2\alpha_2 h^{-1} P_{\bar t} g_2(t)\,, && t \in \vartheta_M;\\
&\big(P_x^2 a_\Delta(u_\Delta)\big)(x,0) = P_x^2 H_0(x)\,, && x \in \overline{\omega}\,.
\end{aligned}\right\} \tag{4.3}$$

Hierbei wurde die Benutzung einheitlicher Mittelungsvorschriften (auch für Gitterrandpunkte) durch vorherige gerade Fortsetzung von u über die Begrenzungsgeraden von Ω_T hinaus ermöglicht.

Mit dem in $L_2(\overline{\omega})$ wirkenden linearen, symmetrischen, nichtnegativen Operator (siehe [14], [15])

$$Av(x) = \begin{cases} -\dfrac{2}{h} v_x(x) + 2\alpha_1 h^{-1} v(x)\,, & x = 0\\ -v_{\bar x x}(x)\,, & x \in \omega\\ +\dfrac{2}{h} v_{\bar x}(x) + 2\alpha_2 h^{-1} v(x)\,, & x = 1 \end{cases} \tag{4.4}$$

werden die Gleichungen (4.3) umgeformt zu

$$\left.\begin{aligned}
&\big((P_x^2 a_\Delta(u_\Delta))_{\bar t} + A P_{\bar t} u_\Delta - P_{\bar t} P_x^2 f(u_\Delta) - g\big)(x,t) = 0\,, && (x,t) \in \overline{\omega}_M;\\
&g(x,t) = \frac{2}{h} \sum_{k=1}^{2} \alpha_k P_{\bar t} g_k(t)\, \delta_{x,k-1}\,, && t \in \vartheta_M\\
&P_x^2 a_\Delta(u_\Delta)(x,0) = P_x^2 H_0(x)\,, && x \in \overline{\omega}\,.
\end{aligned}\right\} \tag{4.5}$$

Das Differenzenschema (2.3) wird mittels (4.4) in der Form

$$\left.\begin{aligned}
&\big(a_\Delta(y)_{\bar t} + Ay - f(y) - g\big)(x,t) = 0\,, && (x,t) \in \overline{\omega}_M\,,\\
&a_\Delta(y)(x,0) = P_x^2 H_0(x)\,, && x \in \overline{\omega}
\end{aligned}\right\}. \tag{4.6}$$

notiert. Für die Gitterfunktionen

$$\left.\begin{aligned}
&z(x,t) = (y - P_{\bar t} u_\Delta)(x,t)\,, && (x,t) \in \overline{\omega}_M\,,\\
&G(x,t) = \big(a_\Delta(y) - P_x^2 a_\Delta(u_\Delta)\big)(x,t)\,, && x \in \overline{\omega}\,, \quad t \in \{0\} \cup \vartheta_M
\end{aligned}\right\} \tag{4.7}$$

¹) Dieser Begriff ist wie in [13], [19] gewählt.

erhält man aus (4.5) und (4.6) das Gleichungssystem

$$\left.\begin{array}{lll}\left(G(z)_{\bar t} + Az - \varphi\right)(x, t) = 0\,, & & (x, t) \in \bar\omega_M\,, \\ \varphi = f(y) - P_{\bar t}P_x^2 f(u_\Delta)\,; & G(x, 0) = 0\,, & x \in \bar\omega\,.\end{array}\right\} \tag{4.8}$$

Bemerkung: Da (4.5) identisch erfüllt ist und (4.6) nach Abschnitt 2 eindeutig lösbar ist, besitzt auch (4.8) eine Lösung.

Lemma 1: *Jede Lösung z des Differenzenschemas* (4.8) *genügt der Beziehung*

$$\left.\begin{array}{l}\sum\limits_{n=1}^{M} \tau\left(G(z)^n + \dfrac{\tau}{2} Az^n, z^n\right) + \dfrac{1}{2}(Av^0, v^0) = \sum\limits_{n=1}^{M} \tau(\varphi^n, v^{n-1})\,, \\ v^n = \begin{cases} 0\,, & n \geqq M \\ \sum\limits_{j=n+1}^{M} \tau z^j\,, & n < M\,.\end{cases}\end{array}\right\} \tag{4.9}$$

Beweis: Das Schema (4.8) wird zunächst in der Form

$$\left(G(z) + \frac{\tau}{2} Az\right)_{\bar t}^{n-1} + \frac{1}{2} A(z^n + z^{n-1}) = \varphi^n\,, \qquad n \geqq 1; \qquad G^0 = G(z^0) = 0$$

notiert und (für jede Zeitschicht) skalar mit τv^{n-1} multipliziert. Es entsteht mit $C(z) = G(z) + 0.5\tau Az$ unter Berücksichtigung von

$$\tau\left(C(z)_t, v\right)^{n-1} = \left(C(z), v\right)^n - \left(C(z), v\right)^{n-1} - \tau\left(C(z), v_{\bar t}\right)^n$$

und nach Summation über n $(1 \leqq n \leqq M)$ die Beziehung

$$\left(C(z), v\right)^M - \sum_{n=1}^{M} \tau\left(C(z), v_{\bar t}\right)^n + \tfrac{1}{2}\sum_{n=1}^{M} \tau\left(A(z^n + z^{n-1}), v^{n-1}\right) = \left(C(z), v\right)^0 + \sum_{n=1}^{M} \tau(\varphi^n, v^{n-1})\,.$$

Beachtet man die aus der Festlegung von v gemäß (4.9) ableitbaren Beziehungen

$$v_{\bar t}^n = -z^n\,, \qquad 1 \leqq n \leqq M; \qquad \frac{\tau}{2}\sum_{n=1}^{M}\left(A(z^n + z^{n-1}), v^{n-1}\right) = \frac{1}{2}(Av, v)^0 + \frac{\tau}{2}(Av, z)^0\,,$$

so folgt wegen $\left(C(z), v\right)^0 - \dfrac{\tau}{2}(Az, v)^0 = \left(G(z), v\right)^0 = 0$ die Behauptung des Lemmas.

Anstatt z als Fehler der Näherungslösung abzuschätzen, wird dies für

$$\overset{*}{z}(x, t) = y(x, t) - u_\Delta(\xi^*, t) = (y - \overset{*}{u}_\Delta)(x, t)\,, \qquad (x, t) \in \bar\omega_M\,, \tag{4.10}$$

angestrebt, wobei $\xi^* = \xi^*(x, t) \in (x - h, x + h)$ so gewählt ist, daß $P_x^2 a_\Delta(u_\Delta)(x, t) = a_\Delta\left(u_\Delta(\xi^*, t)\right)$ gilt (wegen der Stetigkeit von $a_\Delta(u_\Delta)$ und von $u_\Delta(x, t)$ für jedes $t \in \vartheta_M$ existiert jeweils ein solches ξ^*).

Lemma 2: *Mit* (4.10) *ergibt sich für die linke Seite von* (4.9) *die Abschätzung*

$$\left.\begin{array}{l}\sum\limits_{n=1}^{M} \tau\left(G(z) + \dfrac{\tau}{2} Az, z\right)^n + \dfrac{1}{2}(Av, v)^0 \geqq c_0\|\overset{*}{z}\|_{2,\bar\omega_M}^2 - \left(\|a_\Delta(y)\|_{2,\bar\omega_M} + \|a_\Delta(\overset{*}{u}_\Delta)\|_{2,\bar\omega_M}\right)\|\overset{*}{u}_\Delta - P_{\bar t}u_\Delta\|_{2,\bar\omega_M}\,, \\ \|\overset{*}{u}_\Delta - P_{\bar t}u_\Delta\|_{2,\bar\omega_M} \leqq M_5 h + M_6\tau\,, \qquad M_5 = \sqrt{2T}\left(1 + \dfrac{1}{\sqrt{10}}\right)\max\limits_{t\in\vartheta_M}\left\|\dfrac{\partial u_\Delta}{\partial x}(., t)\right\|_{2,\Omega} + \dfrac{1}{\sqrt{5}}\left\|\dfrac{\partial u_\Delta}{\partial x}\right\|_{2,\Omega_T}, \\ M_6 = \dfrac{2}{3}\left\|\dfrac{\partial u_\Delta}{\partial t}\right\|_{2,\Omega_T}\end{array}\right\} \tag{4.11}$$

Beweis: Da A nichtnegativ ist, folgt zunächst mit (4.10)

$$\begin{aligned}\left(G(z) + \frac{\tau}{2} Az, z\right)^n + \frac{1}{2T}(Av, v)^0 &\geqq \left(G(z), z\right)^n = \\ &= \left(a_\Delta(y) - a_\Delta(\overset{*}{u}_\Delta), y - \overset{*}{u}_\Delta\right)^n + \left(a_\Delta(y) - a_\Delta(\overset{*}{u}_\Delta), \overset{*}{u}_\Delta - P_{\bar t}u_\Delta\right)^n \geqq \\ &\geqq c_0\|y^n - \overset{*}{u}{}_\Delta^n\|^2 - \|a(y)^n - a(\overset{*}{u}_\Delta)^n\| \cdot \|\overset{*}{u}{}_\Delta^n - (P_{\bar t}u_\Delta)^n\|\,.\end{aligned}$$

Nach Summation über n entsteht

$$\sum_{n=1}^{M} \tau\left(G(z) + \frac{\tau}{2} Az, z\right)^n + \frac{1}{2}(Av, v)^0 \geqq c_0\|\overset{*}{z}\|_{2,\bar\omega_M}^2 - \|a_\Delta(y) - a_\Delta(\overset{*}{u}_\Delta)\|_{2,\bar\omega_M}\|\overset{*}{u}_\Delta - P_{\bar t}u_\Delta\|_{2,\bar\omega_M}\,.$$

Nun gilt

$$\|\overset{*}{u}_\Delta - P_{\bar t}u_\Delta\|_{2,\bar\omega_M} \leqq \|\overset{*}{u}_\Delta - u_\Delta\|_{2,\bar\omega_M} + \|u_\Delta - P_x^2 u_\Delta\|_{2,\bar\omega_M} + \|P_x^2 u_\Delta - P_{\bar t}P_x^2 u_\Delta\|_{2,\bar\omega_M} + \|P_{\bar t}P_x^2 u_\Delta - P_{\bar t}u_\Delta\|_{2,\bar\omega_M}$$

und die Abschätzung

$$|\overset{*}{u}_\Delta(x,t) - u_\Delta(x,t)| = |u_\Delta(\xi^*,t) - u_\Delta(x,t)| = \left|\int\limits_x^{\xi^*} \frac{\partial u_\Delta}{\partial x}(x')\,dx'\right| \leqq$$

$$\leqq (|\xi^* - x|)^{1/2} \left|\int\limits_x^{\xi^*} \left(\frac{\partial u_\Delta}{\partial x}(x')\right)^2 dx'\right|^{1/2} \leqq \sqrt{h}\left(\int\limits_{x-h}^{x+h}\left(\frac{\partial u_\Delta}{\partial x}\right)^2 dx'\right)^{1/2},$$

d. h.

$$\left.\begin{aligned} &\|\overset{*}{u}_\Delta - u_\Delta\|^2_{2,\bar\omega_M} = \sum_{n=1}^M \tau\|\overset{*}{u}{}^n_\Delta - u^n_\Delta\|^2 \leqq \sum_{n=1}^M \tau\cdot 2h^2 \int_0^1 \left(\frac{\partial u_\Delta}{\partial x}(x',t_n)\right)^2 dx' \leqq M_7^2 h^2\,,\\ &M_7 = \sqrt{2T}\max_{t\in\vartheta_M}\left\|\frac{\partial u_\Delta}{\partial x}(.,t)\right\|_{2,\Omega}.\end{aligned}\right\} \quad (4.12)$$

Auf analoge Weise schätzt man ab (vgl. auch [13]):

$$\left.\begin{aligned} &\|u_\Delta - P_x^2 u_\Delta\|_{2,\bar\omega_M} \leqq M_8 h\,, && M_8 = \left(\frac{T}{5}\right)^{1/2}\max_{t\in\vartheta_M}\left\|\frac{\partial u_\Delta}{\partial x}(.,t)\right\|_{2,\Omega},\\ &\|P_x^2 u_\Delta - P_{\bar t} P_x^2 u_\Delta\|_{2,\bar\omega_M} \leqq M_9\tau\,, && M_9 = \frac{2}{3}\left\|\frac{\partial u_\Delta}{\partial t}\right\|_{2,\Omega_T},\\ &\|P_{\bar t}P_x^2 u_\Delta - P_{\bar t}u_\Delta\|_{2,\bar\omega_M} \leqq M_{10} h\,, && M_{10} = \frac{1}{\sqrt5}\left\|\frac{\partial u_\Delta}{\partial x}\right\|_{2,\Omega_T}.\end{aligned}\right\} \quad (4.13)$$

Damit ergibt sich die Behauptung des Lemmas.

Lemma 3: *Die rechte Seite von (4.9) läßt sich (mit beliebigen $\varepsilon_1 > 0$, $\varepsilon_2 > 0$) abschätzen durch*

$$\begin{aligned}\sum_{n=1}^M \tau(\varphi^n, v^{n-1}) &\leqq (\varepsilon_1 + \varepsilon_2)\,\|\overset{*}{z}\|^2_{2,\bar\omega_M} + \\ &+ \frac{TL^2}{4\varepsilon_1}\sum_{n=1}^M \tau\|\overset{*}{z}\|^2_{2,\bar\omega_n} + \frac{T^2}{4\varepsilon_2}\,\|f(\overset{*}{u}_\Delta) - P_{\bar t}P_x^2 f(u_\Delta)\|^2_{2,\bar\omega_M} + T\|\varphi\|_{2,\bar\omega_M}\,\|\overset{*}{u}_\Delta - P_{\bar t}u_\Delta\|_{2,\bar\omega_M}\,.\end{aligned} \quad (4.14)$$

Beweis: Man stellt zunächst dar mit

$$\overset{*}{v}{}^{n-1} = \sum_{j=n}^M \tau \overset{*}{z}{}^j\,, \qquad n \leqq M\,, \qquad \overset{*}{v}{}^M = 0; \qquad \varphi^n = \overset{(1)}{\varphi}{}^n + \overset{(2)}{\varphi}{}^n\,, \qquad \overset{(1)}{\varphi}{}^n = f(y) - f(\overset{*}{u}_\Delta)\,,$$

$$\overset{(2)}{\varphi}{}^n = f(\overset{*}{u}_\Delta) - P_{\bar t}P_x^2 f(u_\Delta):$$

$$\sum_{n=1}^M \tau(\varphi^n, v^{n-1}) = \sum_{n=1}^M \tau\big((\overset{(1)}{\varphi}{}^n, \overset{*}{v}{}^{n-1}) + (\overset{(2)}{\varphi}{}^n, \overset{*}{v}{}^{n-1}) + (\varphi^n, v^{n-1} - \overset{*}{v}{}^{n-1})\big)\,.$$

Es ist

$$\|\overset{*}{v}{}^{n-1}\| \leqq \Big(\sum_{j=n}^M \tau\Big)^{1/2}\Big(\sum_{j=n}^M \tau\|\overset{*}{z}{}^j\|^2\Big)^{1/2} \leqq \big((M-n+1)\,\tau\big)^{1/2}\,\|\overset{*}{z}\|_{2,\bar\omega_M}\,,$$

d. h.

$$\|\overset{*}{v}\|_{2,\bar\omega_M} \leqq \big(M(M+1)/2\big)^{1/2}\,\tau\|\overset{*}{z}\|_{2,\bar\omega_M} \leqq T\|\overset{*}{z}\|_{2,\bar\omega_M}$$

sowie

$$\|v - \overset{*}{v}\|_{2,\bar\omega_M} \leqq T\|\overset{*}{u}_\Delta - P_{\bar t}u_\Delta\|_{2,\bar\omega_M}\,.$$

Damit wird

$$\sum_{n=1}^M \tau(\varphi^n, v^{n-1}) \leqq \Big|\sum_{n=1}^M \tau(\overset{(1)}{\varphi}{}^n, \overset{*}{v}{}^{n-1})\Big| + T\|\overset{(2)}{\varphi}\|_{2,\bar\omega_M}\,\|\overset{*}{z}\|_{2,\bar\omega_M} + T\|\varphi\|_{2,\bar\omega_M}\,\|\overset{*}{u}_\Delta - P_{\bar t}u_\Delta\|_{2,\bar\omega_M}\,. \quad (4.15)$$

Die weitere Abschätzung des ersten Summanden von (4.15) wird mit Hilfe der Darstellung

$$\overset{(1)}{\varphi}{}^n = F_t^{n-1}\,, \qquad F^n = \sum_{j=1}^n \tau\overset{(1)}{\varphi}{}^j\,, \qquad n \geqq 1\,, \qquad F^0 = 0$$

durchgeführt:

$$\begin{aligned}\Big|\sum_{n=1}^M \tau(\varphi^n, \overset{*}{v}{}^{n-1})\Big| &= \Big|\sum_{n=1}^M \tau\big((F^{n-1}, \overset{*}{v}{}^{n-1})_t - (F^n, \overset{*n}{v_{\bar t}})\big)\Big| = \Big|\sum_{n=1}^M \tau(F^n, \overset{*}{z}{}^n)\Big| \leqq\\ &\leqq \sum_{n=1}^M \tau\|F^n\|\,\|\overset{*}{z}{}^n\| \leqq \Big(\sum_{n=1}^M \tau\,\Big\|\sum_{j=1}^n \tau\overset{(1)}{\varphi}{}^j\Big\|^2\,\sum_{n=1}^M \tau\|\overset{*}{z}{}^n\|^2\Big)^{1/2} \leqq\\ &\leqq \Big(\sum_{n=1}^M \tau\cdot n\tau\,\|\overset{(1)}{\varphi}\|^2_{2,\bar\omega_n}\Big)^{1/2}\,\|\overset{*}{z}\|^2_{2,\bar\omega_M} \leqq \sqrt{T}\,L\,\Big(\sum_{n=1}^M \tau\|\overset{*}{z}\|^2_{2,\bar\omega_n}\Big)^{1/2}\,\|\overset{*}{z}\|_{2,\bar\omega_M} \leqq\\ &\leqq \varepsilon_1\|\overset{*}{z}\|^2_{2,\bar\omega_M} + \frac{TL^2}{4\varepsilon_1}\sum_{n=1}^M \tau\|\overset{*}{z}\|^2_{2,\bar\omega_n}\,.\end{aligned}$$

Dabei wurde zuletzt die Lipschitz-Stetigkeit von f und die ε-Ungleichung benutzt. Aus (4.15) folgt sodann nach nochmaliger Anwendung der ε-Ungleichung (mit beliebigem $\varepsilon_2 > 0$) die Behauptung des Lemmas.

Die Ergebnisse der Lemmata 1 bis 3 können zusammengefaßt werden zu der Beziehung

$$\left(c_0 - \varepsilon_1 - \varepsilon_2 - \frac{TL^2\tau}{4\varepsilon_1}\right) \|\overset{*}{z}\|^2_{2,\bar{\omega}_M} \leqq \frac{TL^2}{4\varepsilon_1} \sum_{n=1}^{M-1} \tau \|\overset{*}{z}\|^2_{2,\bar{\omega}_n} + \frac{T^2}{4\varepsilon_2} \|f(\overset{*}{u}_\Delta) - P_{\bar{t}} P_x^2 f(u_\Delta)\|^2_{2,\bar{\omega}_M} +$$
$$+ (\|a_\Delta(y)\|_{2,\omega_M} + \|a_\Delta(\overset{*}{u}_\Delta)\|_{2,\bar{\omega}_M} + T\|\varphi\|_{2,\bar{\omega}_M}) \|\overset{*}{u}_\Delta - P_{\bar{t}} u_\Delta\|_{2,\bar{\omega}_M} .$$

Wenn die Bedingung

$$\frac{TL^2\tau}{c_0^2} = q < 1 \tag{4.16}$$

erfüllt ist, dann erhält man mit $\varepsilon_1 = \frac{c_0}{2}\sqrt{q}$, $\varepsilon_2 = \frac{c_0}{2}(1 - \sqrt{q})$ die Beziehung

$$\|\overset{*}{z}\|^2_{2,\bar{\omega}_M} \leqq \frac{TL^2}{c_0^2\sqrt{q}(1 - \sqrt{q})} \sum_{n=1}^{M-1} \tau \|\overset{*}{z}\|^2_{2,\bar{\omega}_n} + \frac{T^2}{c_0^2(1 - \sqrt{q})^2} \|f(\overset{*}{u}_\Delta) - P_{\bar{t}} P_x^2 f(u_\Delta)\|^2_{2,\bar{\omega}_M} +$$
$$+ \frac{2}{c_0(1 - \sqrt{q})} (\|a_\Delta(y)\|_{2,\bar{\omega}_M} + \|a_\Delta(\overset{*}{u}_\Delta)\|_{2,\bar{\omega}_M} + T\|\varphi\|_{2,\bar{\omega}_M}) \|\overset{*}{u}_\Delta - P_{\bar{t}} u_\Delta\|_{2,\bar{\omega}_M}$$

und daraus mit Hilfe des Gronwallschen Lemmas (vgl. [14], S. 311) die Abschätzung

$$\left.\begin{aligned} &\|\overset{*}{z}\|^2_{2,\bar{\omega}_M} \leqq M_{11}^2 \|f(\overset{*}{u}_\Delta) - P_{\bar{t}} P_x^2 f(u_\Delta)\|^2_{2,\bar{\omega}_M} + M_{12}^2 \|\overset{*}{u}_\Delta - P_{\bar{t}} u_\Delta\|_{2,\bar{\omega}_M} , \\ &M_{11}^2 = \frac{T^2 M_{13}}{c_0^2(1 - \sqrt{q})^2} , \qquad M_{13} = \exp \frac{T^2L^2}{c_0^2\sqrt{q}(1 - \sqrt{q})} , \qquad M_{12}^2 = \frac{2M_{13}M_{14}}{c_0(1 - \sqrt{q})} , \end{aligned}\right\} \tag{4.17}$$

wobei M_{14} eine Schranke für $\|a_\Delta(y)\|_{2,\bar{\omega}_M} + \|a_\Delta(\overset{*}{u}_\Delta)\|_{2,\bar{\omega}_M} + T\|\varphi\|_{2,\bar{\omega}_M}$ darstellt (diese existiert nach Abschnitt 2 unabhängig von den Parametern h, τ, Δ). Mit

$$\|f(\overset{*}{u}_\Delta) - P_{\bar{t}} P_x^2 f(u_\Delta)\|_{2,\bar{\omega}_M} \leqq$$
$$\leqq L(\|\overset{*}{u}_\Delta - u_\Delta\|_{2,\bar{\omega}_M} + \|u_\Delta - P_x^2 u_\Delta\|_{2,\bar{\omega}_M} + \|P_x^2 u_\Delta - P_{\bar{t}} P_x^2 u_\Delta\|_{2,\bar{\omega}_M}) + \|f(P_{\bar{t}} P_x^2 u_\Delta) - P_{\bar{t}} P_x^2 f(u_\Delta)\|_{2,\bar{\omega}_M} ,$$

der analog zu (4.13) herleitbaren Beziehung

$$\|f(P_{\bar{t}} P_x^2 u_\Delta) - P_{\bar{t}} P_x^2 f(u_\Delta)\|_{2,\bar{\omega}_M} \leqq L(M_{15}h + M_{16}\tau) ,$$
$$M_{15} = 2\left\|\frac{\partial u_\Delta}{\partial x}\right\|_{2,\Omega_T} , \qquad M_{16} = \frac{2}{\sqrt{3}}\left\|\frac{\partial u_\Delta}{\partial t}\right\|_{2,\Omega_T}$$

sowie mit $\|y - u_\Delta\|_{2,\bar{\omega}_M} \leqq \|\overset{*}{u}_\Delta - u_\Delta\|_{2,\bar{\omega}_M} + \|\overset{*}{z}\|_{2,\bar{\omega}_M}$ und den Abschätzungen (4.11), (4.12), (4.13) entsteht aus (4.17) die Ungleichung

$$\|y - u_\Delta\|_{2,\bar{\omega}_M} \leqq M_7 h + M_{11} L\big((M_7 + M_8 + M_{15})\, h + (M_9 + M_{16})\, \tau\big) + M_{12}(M_5 h + M_6 \tau)^{1/2} .$$

Damit ist bewiesen

Satz 2: *Unter den in den Abschnitten* 1 *und* 2 *genannten Voraussetzungen konvergieren die Lösungen* y *des Differenzenschemas* (2.3), (2.1) *für* $h, \tau \to 0$ *gegen die Lösung* u_Δ *der Aufgabe* (1.3), (2.1). *Für gemäß* (4.16) *genügend kleine* τ *gilt die Fehlerabschätzung*

$$\|y - u_\Delta\|_{2,\bar{\omega}_M} \leqq M_{17}(h^{1/2} + \tau^{1/2}) + O(h + \tau) , \qquad M_{17} = M_{12} \max\{M_5^{1/2}, M_6^{1/2}\} . \tag{4.18}$$

Bemerkung: Die in (4.18) eingehenden Konstanten sind gemäß (4.11), (4.17) und den Aussagen von Abschnitt 2 nur von den Daten der Aufgabe (1.2), (1.3) abhängig, so daß mit Satz 2 eine a-priori-Fehlerabschätzung gegeben ist.

Folgerung: *Sei* $\tilde{y}$ *die stückweise konstante* L_2*-Fortsetzung der aus* (2.3) *ermittelten Gitterfunktion* y. *Dann gilt auch in der „kontinuierlichen Norm"*

$$\|y - u_\Delta\|_{2,\Omega_T} \leqq M_{17}(h^{1/2} + \tau^{1/2}) + O(h + \tau) . \tag{4.19}$$

Denn mit $\tilde{u}_\Delta$ als stückweise konstanter L_2-Fortsetzung der Gitterfunktion $u_\Delta(x, t)$, $(x, t) \in \overline{\omega}_M$, hat man zunächst

$$\|\tilde{y} - u_\Delta\|_{2,\Omega_T} \leqq \|\tilde{y} - \tilde{u}_\Delta\|_{2,\Omega_T} + \|\tilde{u}_\Delta - u_\Delta\|_{2,\Omega_T} .$$

Da

$$\|\tilde{u}_\Delta - u_\Delta\|_{2,\Omega_T} \leqq \sqrt{\frac{T}{2}}\, h \max_{0 \leq n \leq M} \left\|\frac{\partial u_\Delta}{\partial x}(., t_n)\right\|_{2,\Omega} + \frac{\tau}{\sqrt{2}}\left\|\frac{\partial u_\Delta}{\partial t}\right\|_{2,\Omega_T}$$

gilt, folgt mit $\|\tilde{y} - \tilde{u}_\Delta\|_{2,\Omega_T} = \|y - u_\Delta\|_{2,\bar{\omega}_M}$ die Behauptung (4.19).

5. Abschließende Bemerkungen

1° Unter Ausnutzung multiplikativer Ungleichungen (vgl. [16]) erhält man die Aussagen

$$\|u_\Delta - u\|_{\infty,\Omega_T} \leqq M_{18}\Delta^{1/16} \tag{5.1}$$

sowie

$$\|y - u_\Delta\|_{\infty,\bar{\omega}_M} \leqq M_{19}(h^{1/8} + \tau^{1/8}),$$

bzw. insgesamt

$$\|y - u\|_{\infty,\bar{\omega}_M} \leqq M_{20}(\Delta^{1/16} + h^{1/8} + \tau^{1/8}).$$

Mit (5.1) wird speziell das Resultat über die gleichmäßige Konvergenz (schlechthin) von u_Δ gegen u (bei glattem $a_\Delta(.)$) aus [7] bestätigt.

2° Das in der Folgerung zum Satz 2 erhaltene Resultat (4.19) korrespondiert in gewisser Weise mit der Fehlerabschätzung für eine implizite Geradenmethode zur Lösung von

$$\frac{\partial a(u)}{\partial t} = \frac{\partial^2 u}{\partial x^2}, \qquad a(\xi) = |\xi|^{1/\gamma}\operatorname{sgn}\xi, \qquad \gamma > 1,$$

die in [20] begründet ist mit einer L_2-Konvergenzordnung von $O(\tau^{1/2+1/2\gamma})$.

3° Die in den Abschnitten 3 und 4 benutzten Methodiken sind für eine größere Aufgabenklasse anwendbar. Dies gilt zunächst für die Randbedingungen (z.B. solche erster Art, wie sie in [2], [3] zugrundegelegt wurden). Der Fall $c_i = c_i(u)$, $c_i(u) \geqq c_0 > 0$, $i = 1, 2$, ist unmittelbar eingeschlossen, ebenso das Auftreten mehrerer Phasenumwandlungen (Mehrphasenaufgaben), da von den Funktionen $a_\Delta^{-1}(.)$, $\Delta \geqq 0$ lediglich die Eigenschaften (3.5) und

$$O(\Delta)\,(\xi' - \xi'')^2 \leqq \left(a_\Delta^{-1}(\xi') - a_\Delta^{-1}(\xi'')\right)(\xi' - \xi'') \leqq c_0^{-1}(\xi' - \xi'')^2 \quad \text{für alle} \quad \xi', \xi'' \in R^1$$

ausgenutzt wurden. Das bedeutet, daß die konkrete Wahl (2.1) für $a_\Delta(\xi)$ nicht zwingend ist, sondern auch jede andere Approximation, die den genannten Voraussetzungen genügt, zu gleichen Resultaten führt. Speziell sind die glatten Approximationen, die auf quasilineare Probleme der Art

$$\tilde{c}(u)\,\frac{\partial u}{\partial t} = \frac{\partial^2 u}{\partial x^2}, \qquad \tilde{c}(\xi) = \frac{\mathrm{d}a_\Delta(\xi)}{\mathrm{d}\xi}$$

führen, eingeschlossen (z. B. [9]). Der Fall temperaturabhängiger Wärmeleitkoeffizienten $k_i(u)$ kann über die Kirchhoff-Transformation erfaßt werden.

4° Der Satz 1 kann leicht für den räumlich n-dimensionalen Fall ($n \geqq 2$) formuliert werden, während eine solche Übertragung von Satz 2 die Stetigkeit von u voraussetzen muß.

Für $n = 1$ ist dies nach Abschnitt 2 gesichert. Ohne die Voraussetzung über die Stetigkeit von u kann durch Übertragung von Resultaten aus [13] für die lineare Wärmeleitungsgleichung auf den nichtlinearen lipschitzstetigen Fall immerhin noch — falls $u \in W_2^{1,1}(\Omega_T)$ —

$$\|y - u_\Delta\|_{2,\bar{\omega}_M} = O\left(\frac{h + \tau}{\Delta}\right) \tag{5.2}$$

abgeleitet werden. Von diesem Zugang her sollten auch bessere Ergebnisse möglich werden, wenn a-priori-Informationen über mes Ω_T^Δ, $\Omega_T^\Delta = \{(x, t) \in \Omega_T : |u_\Delta| \leqq \Delta\}$, bekannt sind.

Das Resultat (5.2) ist übrigens speziell für die Lösung solcher Aufgaben interessant, in denen das Materialgesetz $H = a(u)$ ohnehin einer Lipschitzbedingung mit festem $\Delta > 0$ genügt, ein Übergang $\Delta \to 0$ also irrelevant ist.

5° Zahlreiche neuere Arbeiten empfehlen numerische Methoden (auch implizite) für die „ungeglättete" Aufgabe ($\Delta = 0$); explizite Varianten sind seit längerem bekannt (z. B. [2], [3]). In Anbetracht dessen, daß Satz 2 Abschätzungen unabhängig von Δ liefert und auch die Voraussetzung über die Stetigkeit von $a_\Delta(\xi)$ abgeschwächt werden kann, sollte (für $n = 1$) eine Übertragung der Resultate zumindest auf die Differenzenmethoden von [2], [3] möglich sein.

Literatur

1 Streit, U., Differenzenmethoden für eindimensionale Zwei-Phasen-Stefan-Aufgaben mit nichtkonstanter Temperatur auf der freien Grenze, Diss. A, TH Karl-Marx-Stadt 1980.
2 Каменомостская, С. Л., О задаче Стефана, Мат. сборник 53(95) (1961), 489—514.
3 Atthey, D. R., A finite difference scheme for melting problems, J. Inst. Math. Appl. 13 (1974), 353—366.
4 Jerome, J. W., Nonlinear equations of evolution and a generalized Stefan problem, J. Differential Equ. 26 (1977), 240—261.
5 Ciavaldini, J. F., Analyse numérique d'un problème de Stefan a deux phases par une méthode d'éléments finis, SIAM J. Numer. Anal. 12 (1975), 464—487.
6 Elliot, C. M., On the finite element approximation of an elliptic variational inequality arising from an implicit time discretization of the Stefan problem, IMA J. Numer. Anal. 1 (1981), 115—125.
7 Олейник, О. А., Об одном методе решения общей задачи Стефана, Доклады Акад. Наук СССР 135 (1960), 1054—1057.

8 Jerome, J. W., Existence and approximation of weak solutions of nonlinear Dirichlet problems with discontinuous coefficients, SIAM J. Math. Anal. **9** (1978), 730—742.
9 Будак, Б. М.; Соловьева, Е. Н.; Успенский, А. Б., Разностный метод со сглаживанием коэффициентов для решения задачи Стефана, Ж. вычислит. мат. и мат. физ. **5** (1965), 828—840.
10 Meyer, G. H., Multidimensional Stefan problems, SIAM J. Numer. Anal. **10** (1973), 522—538.
11 Furzeland, R. M., A comparative study of numerical methods for moving boundary problems, J. Inst. Math. Appl. **26** (1980), 411—429.
12 Windisch, G.; Streit, U., Numerische Behandlung von Stefan-Aufgaben, Wiss. Schriftenreihe d. TH Karl-Marx-Stadt, Heft 3 (1981).
13 Streit, U.; Weinelt, W., Untersuchungen zur Konvergenzordnung von Differenzenmethoden für die Approximation schwacher Lösungen. Anwendung auf eine Stefan-Aufgabe, Wiss. Informationen d. Sektion Math. d. TH Karl-Marx-Stadt, Heft 31.
14 Самарский, А. А., Введение в теорию разностных схем, „Наука", Москва 1971.
15 Самарский, А. А., Теория разностных схем, „Наука", Москва 1977.
16 Ладыженская, О. А., Краевые задачи математической физики, „Наука", Москва 1973.
17 Ладыженская, О. А.; Солонников, В. А.; Уральцева, Н. Н., Линейные и квазилинейные уравнения параболического типа, „Наука", Москва 1967.
18 Ortega, J. M.; Rheinboldt, W. C., Iterative Solution of Nonlinear Equations in Several Variables, Academic Press, New York — London 1970.
19 Streit, U.; Weinelt, W., Zur Konvergenzordnung von Differenzenmethoden für die Approximation schwacher Lösungen instationärer Differentialgleichungen, erscheint in „Beiträge zur Numer. Math. **12** (1983).
20 Jerome, J. W., Horizontal Line Analysis of the Multidimensional Porous Medium Equation: Existence, Rate of Convergence, Maximum Principles, Lecture Notes Math. 773, S. 64—82 Springer-Verlag, Berlin-Heidelberg-New York 1980.

Bei der Redaktion eingegangen am 6. 5. 1982

Anschrift: Dr. Uwe Streit, Doz. Dr. Sc. Wilfried Weinelt, Technische Hochschule Karl-Marx-Stadt, Sektion Mathematik, PSF 964, DDR-9010 Karl-Marx-Stadt.

BOOK REVIEWS

Nigul, U. / Engelbrecht, J. (eds.), Nonlinear Deformation Waves, Tallinn, USSR, 1982. Berlin-Heidelberg-New York-Tokyo, Springer-Verlag 1983. XXIII, 453 S., 145 Abb., DM 98,–. US $ 42.30. ISBN 3-540-1226-8 (IUTAM)

Nichtlineare Wellen sind gegenwärtig für viele Forschungsgebiete von aktuellem Interesse. In der Plasmaphysik, der Kontinuumsmechanik, der Biologie und Chemie sowie in der Geophysik spielen sie häufig eine Rolle. Sie führen bei partiellen Differentialgleichungen zu vielfältigen mathematischen Fragestellungen. Der vorliegende Band bringt Vorträge eines internationalen Symposiums, das im August 1982 in Tallinn stattfand. Das Programm gliederte sich in die Themen: 1. nichtlineare mathematische Modelle für Festkörper und Flüssigkeiten, 2. analytische und numerische Methoden für die nichtlineare Wellenausbreitung, 3. inverse Probleme und 4. experimentelle Untersuchungen und Anwendungen.

Die in den Beiträgen behandelten mathematischen Probleme sind in den meisten Fällen aus konkreten Fragestellungen der Hydro- und Gasdynamik, der Elastizitäts- und Plastizitätstheorie, der Akustik oder der Elektrodynamik abgeleitet. Von Solitonlösungen verschiedener Evolutionsgleichungen, insbesondere der Korteweg-de Vries-Gleichung, ist die Rede; Wechselwirkungen von Wellen untereinander und mit materiellen Körpern werden behandelt.

Die 45 Beiträge spiegeln den gegenwärtigen Stand der internationalen Forschung wieder. Neben mathematisch strengen Methoden beobachtet man heuristische Ansätze. Der Inhalt ist in 5 Teile gegliedert: Teil 1: Analytische Methoden und Evolutionsgleichungen, Teil 2: Wellen in plastischen und anderen unelastischen Medien, Teil 3: inverse Probleme, Teil 4: nichtlineare Akustik, Teil 5: kombinierte Felder und Anwendungen.

Dies ist nicht das erste Symposium über nichtlineare Wellen und man kann vermuten, daß weitere folgen werden.

Berlin J. Förste

Paterson, A. R., A First Course in Fluid Dynamics. Cambridge et al., Cambridge University Press 1983. VII, 528 S., £ 30.00 B H/c. US $ 59.50 (ISBN 0521254167); £ 12.50 A P/b. US $ 24.95 (ISBN 0 521 273424 9)

Der vorliegende Band bietet auf ansprechendem Niveau eine umfangreiche Einführung in die Dynamik fluider Systeme, die in erster Linie für Physik- und Mathematikstudenten gedacht ist. Hervorzuheben ist die konzentrierte Darstellung der mathematischen und physikalischen Grundlagen, womit dem Charakter des Buches entsprochen wird.

Es folgt die Behandlung der allgemeinen dynamischen Theorie der Flüssigkeiten und Gase. Diesen Problemkreisen sind die ersten 11 Kapitel des Buches gewidmet, während die zweite Hälfte — Kapitel 12 bis 17 — interessanten Anwendungen vorbehalten ist, die aufgrund der intensiven Vorbereitung recht konzentriert behandelt werden können. Im einzelnen seien aus der grundlegenden Theorie die methodenorientierten Kapitel der Lösungen der Navier-Stokes-Gleichung, die reibungsfreien Strömungen und die Potentialtheorie erwähnt. Es überrascht in diesem Zusammenhang, daß die Einführung des komplexen Potentials erst in den letzten beiden Kapiteln erfolgt, das damit als spezielle Anwendungsform gekennzeichnet wird. Erfreulich dagegen ist die Berücksichtigung thermodynamischer Erscheinungen einschließlich einiger Überlegungen zum Gleichgewicht in der Atmosphäre. Die Nutzung der im ersten Teil des Buches gegebenen Theorie ist sehr konkret und praxisbezogen. Genannt seien die Kapitel über Schallwellen in fluiden Medien — einschließlich einiger Anwendungen auf Musikinstrumente — und das sehr interessante Kapitel über Wasserwellen. Zwei weitere Kapitel sind schließlich den Hochgeschwindigkeitsströmungen und den stationären Oberflächenwellen in Kanälen gewidmet. Speziell wird dabei u. a. auf Unter- und Überschallströmungen, ebene Schockwellen und nichtlineare Effekte — solitäre Wellen — eingegangen.

Die Anschaulichkeit des vorliegenden Buches wird durch eine große Anzahl ausgezeichneter Abbildungen wirkungsvoll erhöht. Eine Fülle interessanter Übungen mit Lösungshinweisen und Lösungen ermöglicht bei konsequenter Nutzung eine gute Selbstkontrolle.

Druck und Ausstattung des Buches sind vorbildlich. Ein klarer Stil erleichtert das Studium des Werkes.

Insgesamt stellt der Band sicher eine Bereicherung der vorhandenen Literatur auf dem Gebiet der Dynamik fluider Systeme dar. Dabei dürfte er aufgrund seiner Anlage auch für interessierte Techniker und Nichtphysiker wertvoll sein, die sich in dieses Gebiet einarbeiten wollen.

Rostock H. Ulbricht

Dallmann, H. / Elster, K.-H., Einführung in die Höhere Mathematik für Naturwissenschaftler und Ingenieure. Band III. Jena, VEB Gustav Fischer Verlag 1983. 632 S., 133 Abb., M 45,–. Ausland 49,–. BN 533 511 1

Der nun vorliegende Band III beschließt das Mathematik-Lehrwerk für Studierende der Natur- und Ingenieurwissenschaften. Er enthält gewöhnliche und partielle Differentialgleichungen, Funktionentheorie, Wahrscheinlichkeitsrechnung und eine Einführung in die mathematische Statistik. Das bereits im Band II im Vergleich mit Büchern ähnlicher Zielstellung herangezogene Niveau bezüglich mathematischer Strenge (siehe auch die Rezension zu Band II, ZAMM 62 (1982) 8, Seite 426) hält auch der Band III. An Besonderheiten erwähnen wir den Stabilitätsbegriff von Ljapunow, einige numerische Verfahren für gewöhnliche Differentialgleichungen erster Ordnung, verallgemeinerte Potenzreihenansätze bei Differentialgleichungen mit Singularitäten (Satz von Fuchs), die für Elektroingenieure nützliche LAPLACE-Transformation, einige Grenzwertsätze und Tafeln für Normalverteilung, χ^2-, t- und F-Verteilung. Die Lektüre des Bandes macht wieder bewußt, daß die derzeit an unseren Hochschulen praktizierte Mathematikausbildung für Ingenieure an einigen Überschriften total vorübergeht, die noch vor wenigen Jahren zum festen Bestand zählten. So werden die drei vom Verlag vorzüglich ausgestatteten Bände dem Studierenden nicht nur eine wertvolle Hilfe bei der Festigung und Vertiefung der Mathematikvorlesung sein, sondern ihm auch das Selbststudium nicht behandelter Themen ermöglichen. Zahlreiche praxisorientierte Beispiele und Übungsaufgaben mit Lösungsangabe werden ihn dabei unterstützen. Aber auch der Lehrende wird nicht zuletzt durch die informativen Beispiele und Modelle zahlreiche Hinweise für die Gestaltung seiner Vorlesung erhalten.

Dresden H. Bialy

Lisek, B. / Hochschild, J., Sequentielle Zuverlässigkeitsprüfung. Leipzig, BSB B. G. Teubner VG 1983. 152 S., M 16,–. BN 666 164 8 (Teubner-Texte zur Mathematik 53)

Das Buch wendet sich an Anwender, die die Zuverlässigkeit von Erzeugnissen prüfen wollen und bereit sind, dabei Methoden der mathematischen Statistik zu benutzen. Mehr als Grundkenntnisse eines naturwissenschaftlichen, technischen oder ökonomisch-statistischen Fachgebietes werden nicht benötigt. Es wird ausschließlich von exponentialverteilten Abständen zwischen den Ausfällen der Erzeugnisse ausgegangen. Die Vielzahl der mit der Anwendbarkeit im Zusammenhang stehenden Probleme und die Ausführlichkeit der Modellbeschreibung sind besonders hervorzuheben. Die Autoren plaudern mit dem Leser über seine Probleme, Hoffnungen und Wünsche bei der Zuverlässigkeitsprüfung. Hoffentlich verkraftet der angestrebte breite Leserkreis die geringe Redundanz insbesondere in Bezug auf die Voraussetzungen und Symbole in der Darstellung. Viele exakte Methoden, Formeln und Faustregeln, suggestive numerische Vergleiche und Tabellen werden geboten. Die Autoren gehen davon aus, daß detaillierte Anwendungen jedoch tiefere Kenntnisse über die statistischen Verfahren bedingen. An wenigen Stellen geht die Ausgewogenheit verloren: Die theoretisch gewagten gestutzten sequentiellen Prüfpläne werden hingenommen, andererseits aber werden den Leserkreis täuschende Bedenken gegen α — β-Test's hervorgehoben, werden Häufigkeitsinterpretationsprobleme Bayesscher Modelle erörtert und gleichzeitig „Gegenbeispiele" unterstellt nach dem Muster „alle Erzeugnisse sind schlecht". Das Buch füllt sehr gut eine Lücke in der Anwender belehrenden Literatur.

Berlin W. Warmuth

ZAMM · Z. Angew. Math. u. Mech. 65 (1985) 12, 49 – 55

NEUMAIER, A.

Existence Regions and Error Bounds for Implicit and Inverse Functions

Dedicated to Prof. KARL NICKEL on the occasion of his sixtieth birthday.

Es werden verschiedene Sätze bewiesen, die die Existenz implizit definierter bzw. inverser Funktionen in vorgeschriebenen Gebieten garantieren. Als Anwendungen werden Fehlerschranken für ungenaue Lösungsabbildungen impliziter Gleichungssysteme und für angenäherte Nullstellen für Gleichungssysteme abgeleitet.

Several theorems are proved which guarantee the existence of implicitly defined functions and inverse functions, respectively, in prescribed domains. As applications, error bounds for inaccurate solution maps of implicit systems of equations and for approximate zeros of systems of equations are derived.

Доказываются некоторые теоремы обеспечивающие существование неявно определенных функций и обратных функций, соответственно, в предписанных областях. В качестве примеров выводятся границы погрешности для неточных преобразований решения неявных систем уравнений и для приближеных корней систем уравнений.

Introduction

Motivated by the problem of numerically constructing error bounds for approximate zeros of systems of equations which may or may not involve free parameters we investigate in this paper constructive conditions which guarantee that a given implicit system of equations $F(x, y) = 0$ allows for all y in a given region $E \subseteq \mathbb{R}^m$ a solution function $x = H(y)$ with values in another given region $D \subseteq \mathbb{R}^n$. The map H may or may not be required to be continuous. As a special case, conditions are sought which guarantee that a given n-dimensional function has an inverse function in a given region $E \subseteq \mathbb{R}^n$ with values in $D \subseteq \mathbb{R}^n$.

In contrast to the well-known local results — the inverse and implicit function theorem — which specify conditions for which there are sufficiently small neighbourhoods D, E of points such that the above holds, there are only a few results concerning larger regions. Namely, a number of conditions, like bounded inverse of the derivative ("Hadamard's theorem"), norm coercivity, or uniform monotonicity guarantee global invertibility of a differentiable function (cf. ORTEGA and RHEINBOLDT [7]) — the case $E = \mathbb{R}^n$, and hidden in exercises of DIEUDONNÉ [2] (Ex. X. 2.6 + 7) some global results are derived for the case of implicit functions. But little seems to have been done in the semilocal case, where D and E are bounded regions of $\mathbb{R}^n$. The present paper is designed to fill this gap.

The organization of the material is as follows. In Section 1 we prove as a basic result a semilocal implicit function theorem. Nonsingular (Fréchet-) derivative and a simple boundary condition suffice to ensure the existence of the solution function in the prescribed domain (Theorem 1). As in DIEUDONNÉ's treatment of the global case, the local implicit function theorem is combined with continuation techniques to obtain the result. A simple transformation then allows to extend the theorem to the case when an initial solution point is not known (Theorem 2). As an application, a computable, realistic error bound for approximate solution functions is derived (see the remarks following Proposition 5).

In Section 2 particular cases of the fixpoint theorems of Brouwer and Leray-Schauder are derived from Theorem 1, indicating that there might be a more general theorem. Indeed, using the methods of degree theory, the differentiability assumption can be considerably weakened, resulting in a theorem (Theorem 3) which contains these fixpoint theorems as particular cases. The price paid is that the resulting solution map need no longer be continuous.

Section 3 is concerned with applications to inverse functions. A semilocal inverse function theorem (Theorem 5) is supplemented by two uniqueness criteria (Propositions 6 and 7). The power of the semilocal inverse function theorem is demonstrated by the derivation of the (global) norm coercivity theorem (Corollary 3) and several criteria for the existence of a zero in a prescribed domain. Suggestions how to verify the boundary conditions involved conclude this section.

The terminology and notation is close to that of ORTEGA and RHEINBOLDT [7]. We refer to this book for background results; e.g. reference to result 5.2.1 of [7] is given as OR 5.2.1. Another useful general reference is SCHWETLICK [10].

1. A semilocal implicit function theorem

In this section our fundamental result, the semilocal implicit function theorem, will be proved and discussed. We give a simple and an extended version of the theorem; the extended version has the special feature that it does not require an initial solution point of the implicit system of equations. Finally, a numerical test is given to verify the validity of the boundary condition. It is shown that this test can be used to bound the error in approximate solution maps of implicit functions.

Theorem 1 (Semilocal implicit function theorem): *Let D be an open and bounded subset of $\mathbb{R}^n$, and let E be a simply connected Hausdorff space. Let $F: \overline{D} \times E \to \mathbb{R}^n$ be continuous, and suppose that the following two conditions hold:*

(H1) $\partial_1 F(x, y)$ *exists for* $x \in D$, $y \in E$, *is continuous and nonsingular.*

(H2) $F(x, y) \neq 0$ *for* $x \in \partial D$, $y \in E$.

Then for given $(x^0, y^0) \in D \times E$ *with* $F(x^0, y^0) = 0$, *there is a unique continuous map* $H: E \to D$ *with*

$$Hy^0 = x^0\,, \qquad F(Hy, y) = 0 \quad \text{for} \quad y \in E\,. \tag{1), (2}$$

Here we call the space E *simply connected* if (i) any two points $y^0, y^1 \in E$ can be joined by a (continuous) path $q: [0, 1] \to E$ with $q(0) = y^0$, $q(1) = y^1$, and (ii) if $q = q_0$ and $q = q_1$ are two such paths then there exists a continuous (deformation) map $q: [0, 1] \times [0, 1] \to E$ such that $q(s, 0) = y^0$, $q(s, 1) = y^1$, $q(0, t) = q_0(t)$, $q(1, t) = q_1(t)$ for all $s, t \in [0,1]$. In applications, E often is an interval or a subset of $\mathbb{R}^p$ with the induced topology.

The proof of Theorem 1 is split into several propositions. Proposition 1 is the well-known local implicit function theorem; Proposition 2 is a slight extension for the case when there are several solutions of $F(x, y) = 0$ at a point $y = y^0$. Proposition 3 contains the core of the argument, essentially treating the case where E is the unit square. As an immediate consequence, continuation along arbitrary paths is possible (Proposition 4). Finally, the simple connectedness of E allows to combine Propositions 3 and 4 to prove the theorem.

In the statements of the following propositions we always assume that the hypothesis of Theorem 1 holds.

Proposition 1: *If* $(x^1, y^1) \in D \times E$ *satisfies* $F(x^1, y^1) = 0$ *then there are open neighbourhoods* $D_1 \subseteq D$ *of* x^1 *and* $E_1 \subseteq E$ *of* y^1 *such that the equation* $F(x, y) = 0$ *has for all* $y \in E_1$ *a unique solution* $x = H_1 y \in D_1$. *Moreover, the map* $H_1: E_1 \to D_1$ *defined in this way is continuous.*

Proof: The proof of OR 5.2.4 applies with only trivial modifications. □

Proposition 2: *For* $y^1 \in E$, *the set*

$$X := \{x \in \bar{D} \mid F(x, y^1) = 0\}$$

is finite, and there are open, disjoint neighbourhoods $D(x) \subseteq D$ *of* $x \in X$ *and a closed neighbourhood* $E_1 \subseteq E$ *of* y^1 *such that the equation* $F(\xi, y) = 0$ *has for all* $y \in E_1$ *and all* $x \in X$ *a unique solution* $\xi = H_x y \in D(x)$. *Moreover, the maps* $H_x: E_1 \to D(x)$ *defined in this way are continuous.*

Proof. By (H2), X is a subset of D; hence by Proposition 1 there are (for $x \in X$) open neighbourhoods $D_1(x) \subseteq D$ of x and $E_1(x) \subseteq E$ of y^1 such that the equation $F(\xi, y) = 0$ has for all $y \in E_1(x)$ a unique solution $\xi = H_x y \in D_1(x)$, and $H_x: E_1(x) \to D_1(x)$ is continuous. In particular, for $x \in X$, the set $D_1(x)$ contains no point of X distinct from x. Therefore, X has no accumulation points and since X is closed (F continuous) and bounded (D bounded), X is finite. Hence $\mu := \min \{\|x - x'\| \mid x, x' \in X, x \neq x'\}$ is a positive number, the sets $D(x) := \{\xi \in D_1(x) \mid \|\xi - x\| < \mu/2\}$ are open neighbourhoods of $x \in X$, and $D(x) \cap D(x') = \emptyset$ for $x, x' \in X$, $x \neq x'$. Since X is finite and the H_x are continuous, the intersection E_1' of the sets $\{y \in E_1(x) \mid H_x y \in D(x)\}$, $x \in X$, is a neighbourhood of y^1. If we restrict H_x to a closed neighbourhood E_1 of y^1 contained in E_1', the assertion follows. □

Proposition 3: *If* $(x^0, y^0) \in D \times E$ *satisfies* $F(x^0, y^0) = 0$, *then for every continuous map* $q: [0, 1] \times [0,1] \to E$ *with*

$$q(s, 0) = y^0 \quad \text{for} \quad s \in [0, 1]\,, \tag{3}$$

there is a unique continuous map $p: [0, 1] \times [0, 1] \to D$ *with*

$$p(s, 0) = x^0\,, \qquad F(p(s, t), q(s, t)) = 0 \tag{4), (5}$$

for all $s, t \in [0, 1]$.

Proof: We write $J = [0, 1]$, and denote by Δ a subset of $J \times J$, maximal with respect to the properties

(i) $(s, t) \in \Delta\,, \quad 0 \leqq u \leqq t \Rightarrow (s, u) \in \Delta\,,$

(ii) there is a unique continuous map $p: \Delta \to D$ such that (4) and (5) hold for all $(s, t) \in \Delta$.

Since (3) holds, $(s, 0) \in \Delta$ for all $s \in J$. In order to show that $\Delta = J \times J$ we define for $s \in J$ the number

$$t_s := \sup \{t \in J \mid (s, t) \in \Delta\}\,.$$

Since J is compact there is a number $\sigma \in J$ such that $\tau = t_\sigma$ is minimal. We now put

$$X = \{x \in \bar{D} \mid F(x, q(\sigma, \tau)) = 0\}\,.$$

By Proposition 2 there are disjoint, open neighbourhoods $D(x) \subseteq D$ of $x \in X$ and a closed neighbourhood $E_1 \subseteq E$ of $q(\sigma, \tau)$ such that the equation $F(\xi, y) = 0$ has for all $y \in E_1$ a unique solution $\xi = H_x y \in D(x)$, and the maps H_x are continuous. By continuity of p, the sets

$$\Delta(x) := \{(s, t) \in \Delta \mid p(s, t) \in D(x)\} \tag{6}$$

are disjoint and open in Δ. We now put

$$U(\varepsilon) := \{(s, t) \in J \times J \mid |s - \sigma| \leqq \varepsilon, |t - \tau| \leqq \varepsilon\}\,,$$

and show that there is $\delta > 0$ with

$$\Delta \cap U(\delta) \subseteq \Delta_0 := \bigcup_{x \in X} \Delta(x)\,. \tag{7}$$

For if this is not the case, there is a sequence $(s_i, t_i) \in \Delta \cap U(1/i)$, $i = 1, 2, \ldots$, such that $(s_i, t_i) \notin \Delta_0$, and by construction of $U(\varepsilon)$ we have $\lim_{i \to \infty} (s_i, t_i) = (\sigma, \tau)$. Since D is bounded, the sequence $p(s_i, t_i)$ has at least one accumulation point $x \in \bar{D}$. By continuity of F and q we have $F(x, q(\sigma, \tau)) = 0$, hence $x \in X$. By continuity of p, this implies $p(s_i, t_i) \in \Delta(x)$, hence $(s_i, t_i) \in \Delta(x) \subseteq \Delta_0$ for infinitely many values of i, a contradiction. Therefore, (7) holds for some $\delta > 0$.

By minimality of τ, $\Delta \cap U(\delta)$ is connected (two points (s, t), (s', t') are joined by line segments joining (s, t) — (s, t_0) — (s', t_0) — (s', t'), where $t_0 = \max (0, \tau - \delta)$, and this path is in $\Delta \cap U(\delta)$ if its endpoints are). Since the $\Delta(x)$ are disjoint and open in Δ, $\Delta \cap U(\delta)$ is contained in one of the $\Delta(x)$, say, in $\Delta(\bar{x})$, $\bar{x} \in X$. By (6),

$$p(s, t) \in D(\bar{x}) \quad \text{for} \quad (s, t) \in \Delta \cap U(\delta)\,. \tag{8}$$

Denote by $\bar{\tau}$ the largest $t \in J$ with $q(\sigma, t) \in E_1$. By continuity of q,

$$\text{either} \quad \bar{\tau} > \tau, \quad \text{or} \quad \bar{\tau} = \tau = 1.$$

By (8) and the construction of $D(\bar{x})$, E_1, and $H_{\bar{x}}$, the unique continuous continuation of p along the line segment $s = \sigma, \tau \leqq t \leqq \bar{\tau}$ is given by $p(s, t) := H_{\bar{x}} q(s, t)$. Hence the maximality of Δ implies $\bar{\tau} = \tau = 1$, $(\sigma, 1) = (\sigma, \bar{\tau}) \in \Delta$. Therefore, $\Delta = J \times J$, and the assertion is proved. □

Proposition 4: *If $(x^0, y^0) \in D \times E$ satisfies $F(x^0, y^0) = 0$ then for every continuous map q_0: $[0, 1] \to E$ with $q_0(0) = y^0$ there is a unique continuous map p_0: $[0, 1] \to D$ satisfying*

$$p_0(0) = x^0, \qquad F(p_0(t), q_0(t)) = 0 \quad \textit{for all} \quad t \in [0, 1]. \tag{4a), (5a}$$

Proof: We apply Proposition 3 to the map q defined by $q(s, t) := q_0(t)$ for $s, t \in [0, 1]$. Then (4a) and (5a) hold for the map p_0 defined by $p_0(t) = p(0, t)$ for $t \in [0, 1]$. Conversely, if (4a) and (5a) hold then the map p' defined by $p'(s, t) = p_0(t)$ for $s, t \in [0,1]$ is continuous and satisfies (4) and (5) for $s, t \in [0, 1]$ (with p replaced by p'). Hence $p' = p$; in particular $p_0(t) = p'(0, t) = p(0, t)$ for $t \in [0, 1]$ and p_0 is unique. □

Proof of Theorem 1: Write again $J = [0, 1]$. Denote for arbitrary paths q_0: $J \to E$ with $q_0(0) = y^0$ by $x(q_0)$ the endpoint $p_0(1)$ of the path p_0 defined by Proposition 4. We show that $x(q_0)$ depends only on the endpoint $y^1 = q_0(1)$ of q_0.

For, suppose that q_1: $J \to E$ is another path with $q_1(0) = y^0$, $q_1(1) = y^1$. Since E is simply connected there is a continuous map q: $J \times J \to E$ such that

$$q(s, 0) = y^0, \quad q(s, 1) = y^1, \quad \text{for} \quad s \in J; \quad q(0, t) = q_0(t), \quad q(1, t) = q_1(t) \quad \text{for} \quad t \in J.$$

By Proposition 3 there is a unique continuous map p: $J \times J \to D$ satisfying (4) and (5). By Proposition 4, $p_0(t)$ coincides with $p(0, t)$; in particular,

$$x^1 := x(q_0) = p_0(1) = p(0, 1).$$

Now the set $\Delta := \{s \in J \mid p(s, 1) = x^1\}$ is closed and nonempty; hence $\sigma := \sup \Delta \in \Delta$. By Proposition 1 there is a neighbourhood D_1 of $p(\sigma, 1) = x^1$ such that $x = x^1$ is the only solution of $F(x, y^1) = 0$ in D_1. If $\sigma < 1$ the continuity of p implies the existence of $s \in (\sigma, 1)$ such that $p(s, 1) \in D_1$. Now $F(p(s, 1), y^1) = F(p(s, 1), q(s, 1)) = 0$, hence $p(s, 1) = x^1$ by construction of D_1. But then $s \in \Delta$, contradicting the maximality of σ. Therefore $\sigma = 1$; in particular, $p(1, 1) = x^1 = x(q_0)$. But by Proposition 4, $p_1(t)$ coincides with $p(1, t)$; hence $x(q_1) = p_1(1) = p(1, 1) = x(q_0)$. Therefore, $x(q_0)$ indeed only depends on the endpoint of q_0.

This allows us to define a map H: $E \to D$ by the rule $Hy := x(q_0)$ where q_0: $J \to E$ is an arbitrary path with $q_0(0) = y^0$, $q_0(1) = y$. This map satisfies (1) and (2), and is continuous by Proposition 1. Conversely, if H' : $E \to D$ is a continuous map satisfying (1) and (2) (with H replaced by H'), and if q_0: $J \to E$ is a path with $q_0(0) = y^0$, $q_0(1) = y$ then the path p_0 defined by $p_0(t) := H' q_0(t)$ for $t \in J$ is (by Proposition 4) the unique path satisfying (4a) and (5a). Therefore $H'y = H'q_0(1) = p_0(1) = x(q_0) = Hy$, and H is unique. □

Remarks:

1. A similar result for the case when $D = \mathbb{R}^n$ is contained in DIEUDONNÉ [2], Ex. X.2.6, with a similar proof. For our applications to error bounds, however, it is relevant that D can be chosen as a small bounded region.

2. The difficulty in the proof of Theorem 1 is to show the existence of a continuous map H solving (1) and (2). To show the existence of some solution $x \in D$ with $F(x, y) = 0$ only one can proceed very simply:

Suppose $y \in E$. Since E is path connected, there is a continuous map p: $[0, 1] \to E$ such that $p(0) = y^0$, $p(1) = y$. The set

$$\Delta := \{t \in [0, 1] \mid F(x, p(t)) = 0 \quad \text{for some} \quad x \in D\}$$

contains zero; hence $\tau := \sup \Delta$ is defined and there is a sequence of numbers $t_i \in \Delta$ with limit τ. By definition of Δ, there are $x_i \in D$ with $F(x_i, p(t_i)) = 0$. Since D is bounded, there is a convergent subsequence x_{i_l}, $l = 1, 2, \ldots$, of the sequence x_i. The limits $x^1 = \lim x_{i_l}$, $y^1 = \lim p(t_{i_l}) = p(\tau)$ satisfy $F(x^1, y^1) = 0$ and by (H2), $x^1 \in D$.

If $\tau < 1$ then by the local implicit function theorem (Proposition 1) the equation $F(x, p(t))$ has a solution x for $p(t)$ in a neighbourhood of $y^1 = p(\tau)$, hence (by continuity of p) for some $t \in (\tau, 1)$, contradicting the maximality of τ. Therefore $\tau = 1$, $y^1 = p(\tau) = y$, and $x = x^1$ is a solution of $F(x, y) = 0$.

3. The differentiability condition (H1) is only used to prove Proposition 1. Therefore, (H1) may be replaced by the statement of Proposition 1.

4. If Theorem 1 shall be applied numerically, two problems arise. Namely, due to finite precision arithmetic, it is often the case that in place of an initial pair (x^0, y^0) with $f(x_0, y_0) = 0$ only a pair with $F(x^0, y^0)$ "small" is available. Moreover, a practical way is needed to verify the boundary condition (H2). In the remainder of this section we show how to handle both problems.

Theorem 2 (Semilocal implicit function theorem — extended form): *Let D be an open and bounded subset of $\mathbb{R}^n$, and let E be a simply connected Hausdorff space. Let F: $\bar{D} \times E \to \mathbb{R}^n$ be continuous, and suppose that for some point $(x^0, y^0) \in D \times E$ and some continuous map Δ: $E \to \mathrm{R}^n$ with $\Delta y^0 = F(x^0, y^0)$, the following two conditions hold:*

(H1) $\partial_1 F(x, y)$ *exists for* $x \in D$, $y \in E$, *is continuous and nonsingular.*

(H2*) $F(x, y) \neq t \Delta y$ for $x \in \partial D$, $y \in E$, $t \in [0, 1]$.

Then there is a continuous map H: $E \to D$ with $F(Hy, y) = 0$ for all $y \in E$.

Proof: Define $E^* := E \times [0, 1]$. Then E^* is again simply connected, the map F^*: $\bar{D} \times E^* \to \mathbb{R}^n$ defined by $F^*(x, y, t) := F(x, y) - t \Delta y$ is continuous, and $\partial_1 F^*(x, y, t) = \partial_1 F(x, y)$ for $(x, y, t) \in D \times E^*$. Moreover $F^*(x^0, y^0, 1) = 0$, and $F^*(x, y, t) \neq 0$ for $x \in \partial D$, $(y, t) \in E^*$ by (H2*). Hence by Theorem 1, there is a continuous map H^*: $E^* \to D$ such that $H^*(y^0, 1) = x^0$ and $F^*(H^*(y, t), y, t) = 0$ for all $(y, t) \in E^*$. Therefore the map H: $E \to D$ defined by $Hy := H^*(y, 0)$ is continuous and satisfies $F(Hy, y) = F^*(Hy, y, 0) = F^*(H^*(y, 0), y, 0) = 0$ for all $y \in E$. □

Note that the conclusion has no reference to (x^0, y^0) whence uniqueness cannot be guaranteed.

The following computable form of the boundary condition (H2*) is similar in spirit to an existence condition for zeros of equations based on Brouwer's fixpoint theorem given in NEUMAIER [5]. In that paper it is also discussed how the required constants can be computed efficiently. The norm in the following may be any monotone vector norm; absolute value and inequalities are meant componentwise.

Proposition 5: *Let D be an open bounded subset of $\mathbb{R}^n$, let E be a connected Hausdorff space, and let $F: \bar{D} \times E \to \mathbb{R}^n$ and $H_0: E \to D$ be continuous functions. For some nonsingular $n \times n$-matrix A, define*

$$\delta_y := A^{-1}(\Delta y) \quad \text{for} \quad y \in E \tag{9}$$

and suppose that there is a number $\varkappa > 1$ such that for all $y \in E$,

$$D_y := \{x \in \mathbb{R}^n \mid \|x - H_0 y\| \leqq \varkappa \|\delta_y\|\} \subseteq D .$$

Let c be a nonnegative vector such that

$$|F(x, y) - \Delta y - A(x - H_0 y)| \leqq c \quad \text{for} \quad (x, y) \in \bar{D} \times E . \tag{10}$$

If the vector

$$b := |A^{-1}| \, c$$

satisfies the inequality

$$\|b\| < (\varkappa - 1) \, \|\delta_y\| \quad \text{for all} \quad y \in E$$

then the boundary condition (H2*) *is satisfied.*

Proof: Suppose that $F(x, y) = t \, \Delta y$ for some $x \in \bar{D}$, $y \in E$, $t \in [0, 1]$. With $x^0 := H_0 y$ we then have by (9) and (10),

$$|x - x^0| \leqq |x - x^0 - (t - 1) \, \delta_y| + (1 - t) \, |\delta_y| \leqq |A^{-1}| \, |(t - 1) \, A\delta_y - A(x - x^0)| + |\delta_y|$$
$$\leqq |A^{-1}| \, |F(x, y) - \Delta y - A(x - x^0)| + |\delta_y| \leqq |A^{-1}| \, c + |\delta_y| = b + |\delta_y|$$

whence $\|x - x^0\| \leqq \|b\| + \|\delta_y\| \leqq \varkappa \, \|\delta_y\|$. Therefore $x \in D_y \subseteq D$, and the boundary condition (H2*) follows. □

Remarks:

1. In order to make the vector c small we may proceed as follows: For A we take an approximation of $\partial_1 F(x^0, y^0)$. For H_0 we take an approximation of the expected solution map; if none is known, a suitable substitute may be

$$H_0 y := x^0 = \text{const}, \quad \text{or} \quad H_0 y := x^0 - A^{-1} F(x^0, y) .$$

Then Δ may be chosen as

$$\Delta y := F(H_0 y, y), \quad \text{or} \quad \Delta y := F(x^0, y^0) = \text{const.}$$

2. The proof of Proposition 5 (case $t = 0$) shows that the solution map H satisfies

$$Hy \in D_y \quad \text{for all} \quad y \in E .$$

Therefore, Theorem 2 combined with Proposition 5 and the preceding remarks amounts to an error bound for approximate solution maps of $F(x, y) = 0$, $y \in E$. The discussion in Neumaier [5] shows that for $\varkappa = 1.5$ and $\Delta y = F(H_0 y, y)$, the bounds obtained will overestimate the true errors by a factor of at most three, and hence be quite realistic.

2. Special cases and a weaker hypothesis

We begin this section with two corollaries of Theorem 1, namely special cases of the fixpoint theorems of Brouwer and Leray-Schauder.

Corollary 1: *Let D be a nonempty, convex, open, and bounded subset of $\mathbb{R}^n$, and let $\Phi: \bar{D} \to \bar{D}$ be a continuously differentiable map. If for all $x \in D$ none of the real eigenvalues of $\Phi'(x)$ exceeds 1 then Φ has a fixpoint in $\bar{D}$.*

Proof: Put $E := [0, 1)$, $y^0 := 0$, and, for a fixed $x^0 \in D$, define

$$F(x, y) := y(\Phi x - x^0) + x^0 - x, \qquad x \in \bar{D}, \qquad y \in [0, 1) .$$

Then $\partial_1 F(x, y) = y\Phi'(x) - I$ is continuous and nonsingular and $F(x^0, y^0) = 0$. Moreover, $F(x, y) = 0$ implies that $x = y(\Phi x) + (1 - y) \, x^0$ is a convex combination of $x^0 \in D$ and $\Phi x \in \bar{D}$, and since $y \neq 1$ we have $x \in D$. Therefore (H2) is satisfied. By Theorem 1 there is now a continuous map $H: [0, 1) \to D$ such that $F(Hy, y) = 0$ for all $y \in [0, 1)$. Clearly, for $y \to 1$, Hy has at least one accumulation point $x^* \in \bar{D}$, and we have $F(x^*, 1) = 0$, i.e. $\Phi x^* = x^*$. □

Corollary 2: *Let D be an open and bounded subset of $\mathbb{R}^n$, and let $\Phi: \bar{D} \to \mathbb{R}^n$ be a continuously differentiable map.* If $0 \in D$,

$$\Phi x \neq \lambda x \quad \text{for all} \quad x \in \partial D, \lambda > 1 ,$$

and if for all $x \in D$ none of the real eigenvalues of $\Phi'(x)$ exceeds 1 then Φ has a fixpoint in $\bar{D}$.

Proof: Put $E := [0, 1)$, $y^0 := 0$, and define

$$F(x, y) := y(\Phi x) - x, \qquad x \in \bar{D}, \qquad y \in [0, 1) .$$

Then $\partial_1 F(x, y) = y\Phi'(x) - I$ is continuous and nonsingular, and $F(x^0, y^0) = 0$. Moreover, $F(x, y) = 0$ implies $x = 0$ (if $y = 0$) and $\Phi(x) = \lambda x$, $\lambda = y^{-1} > 1$ (if $y > 0$). Hence by assumption, $x \notin \partial D$. Therefore, (H2) holds, and as before, a fixpoint of Φ exists. □

In fact, both the eigenvalue assumption and the differentiability assumption can be dropped; both results hold for any continuous Φ. This is the contents of the famous fixpoint theorems of Brouwer (cf. OR6. 3.2) and Leray-Schauder (cf. OR6.3.3). This suggests that by weakening the differentiability assumptions of Theorem 1 a stronger existence theorem can be proved which contains the Brouwer and Leray-Schauder fixpoint theorems as special cases. Indeed, this is possible, but the deeper methods of degree theory are required.

Theorem 3: *Let D be an open bounded subset of $\mathbb{R}^n$, and let E be a connected Hausdorff space. Let $F: \bar{D} \times E \to \mathbb{R}^n$ be continuous, and suppose that for some pair $(x^0, y^0) \in D \times E$ with $F(x^0, y^0) = 0$, the following two conditions hold.*

(H1*) $F_0(x) := F(x, y^0)$ *has in D a continuous, nonsingular derivative .*

(H2) $F(x, y) \neq 0$ *for* $x \in \partial D$, $y \in E$.

Then the equation $F(x, y) = 0$ has for all $y \in E$ at least one solution $x \in D$.

Proof: We shall use degree theory as discussed in Chapter 6 of ORTEGA and RHEINBOLDT [7]. The map $F_0: x \to F(x, y^0)$ satisfies the assumption of OR6.1.2 (with obvious change of notation); hence by OR6.1.5,

$$\deg (F_0, D, 0) = \sum \operatorname{sgn} \det F_0'(x) ,$$

where the sum extends over all zeros $x \in D$ of F_0. Since $F_0(x^0) = F(x^0, y^0) = 0$, the sum is not empty, and since $\det F_0'(x)$ is continuous and nonzero in D it has constant sign. Therefore $\deg (F_0, D, 0) \neq 0$. Now for arbitrary $y \in E$, there is a path $q: [0, 1] \to E$ such that $q(0) = y^0$, $q(1) = y$, and the map $H: \bar{D} \times [0, 1] \to \mathbb{R}^n$ defined by $H(x, t) := F(x, q(t))$ satisfies the assumption of OR6.2.2. Hence

$$\deg (F(\cdot, y), D, 0) = \deg (H(\cdot, 1), D, 0) = \deg (H(\cdot, 0), D, 0) = \deg (F_0, D, 0) \neq 0 ,$$

and by OR6.3.1, the equation $F(x, y) = 0$ has a solution $x \in D$. □

Remark: If we now repeat the arguments leading to Corollary 1 and 2 but use Theorem 3 instead of Theorem 1 we just obtain the fixpoint theorems of Brouwer and Leray-Schauder; the differentiability assumption in (H1*) becomes trivial since $F_0(x)$ has now the form $x^0 - x$ and $-x$, respectively.

To free ourselves from the apriori knowledge of an initial pair (x^0, y^0) with $F(x^0, y^0) = 0$ we may proceed as in Theorem 2 and obtain (the details are left to the reader):

Theorem 4: *Let D be an open and bounded subset of $\mathbb{R}^n$, and let E be a connected Haussdorff space. Let $F: \bar{D} \times E \to \mathbb{R}^n$ be continuous, and suppose that for some pair $(x^0, y^0) \in D \times E$ and some continuous map $\Delta: E \to \mathbb{R}^n$ with $\Delta y^0 = F(x^0, y^0)$, the following two conditions hold:*

(H1*) *$F_0(x) := F(x, y^0)$ has in D a continuous, nonsingular derivative,*

(H2*) $F(x, y) \neq t \Delta y$ for $x \in \partial D$, $y \in E$, $t \in [0, 1]$.

Then the equation $F(x, y) = 0$ has for all $y \in E$ at least one solution $x \in D$. □

Remarks:

1. Since Proposition 5 has no differentiability assumption, it can be used even in this more general situation to verify the boundary condition (H2*).
2. Under the assumptions of Theorem 3 or Theorem 4, the set-valued map $H: E \to \mathfrak{P}(D)$ with

$$\{Hy := x \in D \mid F(x, y) = 0\}$$

is upper semicontinuous; i.e. if $(x^i, y^i) \in D \times E$, $i = 1, 2, \ldots$, is a convergent sequence with limit $(x, y) \in D \times E$ then

$$x^i \in H(y^i) , \quad i = 1, 2, \ldots \Rightarrow x \in H(y) .$$

3. Inverse functions

In this section we apply the preceding results to the problem of inverting a map $F: \bar{D} \subseteq \mathbb{R}^n \to \mathbb{R}^n$. Moreover, we point our several existence tests for zeros of such a map some of which can be easily verified computationally.

Theorem 5 (semilocal inverse function theorem): *Let D be an open and bounded subset of $\mathbb{R}^n$, and let $F: \bar{D} \to \mathbb{R}^n$ be a continuous function which has in D a nonsingular, continuous derivative. Let E be a simply connected subset of $\mathbb{R}^n$ such that*

$$F(x) \notin E \quad \text{for} \quad x \in \partial D .$$

Then for given $x^0 \in D$ with $F(x^0) \in E$ there is a unique continuous map $F^{-1}: E \to D$ such that

$$F^{-1}(F(x^0)) = x^0 , \qquad F(F^{-1}(y)) = y \quad \text{for} \quad y \in E .$$

Moreover, F^{-1} is continuously differentiable in E, and for $y \in E$ we have

$$x = F^{-1}(y) \Rightarrow (F^{-1})'(y) = F'(x)^{-1} .$$

Proof: Existence and uniqueness follows from Theorem 1, applied with $y^0 := F(x^0)$ to $F(x) - y$ in place of $F(x, y)$. Differentiability and the formula for the derivative follows from OR5.2.1. □

Remark: For a similar result, in which the assumption "E simply connected" is replaced by the assumption "$E = F(D)$", see POURCIAU [9].

In general, the uniqueness of F^{-1} with $F^{-1}(F(x^0)) = x^0$ need not imply that $F(x) = y \in E$ has a unique solution $x \in D$; cf. OR, E5.3.1. But since we know already the existence of some continuous inverse function, two uniqueness criteria are available; both use the same hypothesis

(I) *Let D^* be a connected subset of $\mathbb{R}^n$, and let $F: D^* \to \mathbb{R}^n$ be a continuous function which has in D^* a nonsingular, continuous derivative. Suppose that on a subset E^* of $\mathbb{R}^n$, F has a continuous inverse $F^{-1}: E^* \to D^*$.*

Proposition 6: *Under the hypothesis* (I), *suppose that*

$$F(x) \in E^* \quad \text{for all} \quad x \in D^* .$$

Then for all $x \in D^*$ we have

$$F(x) = y \Leftrightarrow x = F^{-1}(y) .$$

Proof: Let $x^1 \in D^*$ and put $y^1 := F(x^1)$. Since D^* is connected there is a path $x: [0, 1] \to D^*$ such that $x(0) = x^0$, $x(1) = x^1$. Put $\bar{x}(t) := F^{-1}(F(x(t)))$, and define $\Delta := \{t \in [0, 1] \mid x(t) = \bar{x}(t)\}$. Then Δ is closed and $0 \in \Delta$; in particular $\bar{t} := \sup \Delta \in \Delta$. By Proposition 1 (applied with $F(x) - y$ in place of $F(x, y)$) there are neighbourhoods $D_0 \subseteq D^*$ of $x(\bar{t})$ and $E_0 \subseteq E^*$ of $F(x(\bar{t}))$ such that for $y \in E_0$, the equation $F(x) = y$ has a unique solution $x \in D_0$. If $\bar{t} < 1$ then by continuity of x, $\bar{x}$, and F, there is a $t' \in (\bar{t}, 1)$ such that $x(t')$, $\bar{x}(t') \in D_0$ and $F(x(t')) = F(\bar{x}(t')) \in E_0$; hence $x(t') = \bar{x}(t')$, $t' \in \Delta$, $t' = \sup \Delta \geqq t'$, contradiction. Therefore $\bar{t} = 1$, whence $x^1 = x(1) = \bar{x}(1) = F^{-1}(F(x^1)) = F^{-1}(y^1)$. □

Proposition 7: *Under the hypothesis* (I), *suppose that D^* is convex, and suppose that $\mathfrak{M}$ is a convex set of nonsingular $n \times n$-matrices such that*

$$F'(x) \in \mathfrak{M} \quad \text{for all} \quad x \in D^* .$$

Then

$$x, \bar{x} \in D^* , \qquad F(x) = F(\bar{x}) \Rightarrow x = \bar{x} .$$

Proof: Suppose that $x, \bar{x} \in D^*$ and $F(x) = F(\bar{x})$. Define $y(t) := F(x + t(\bar{x} - x))$. Then $y(0) = y(1)$ and $y'(t) = F'(x + t(\bar{x} - x)) (\bar{x} - x)$. Therefore,

$$0 = y(1) - y(0) = \int_0^1 y'(t) \, dt = \int_0^1 F'(x + t(\bar{x} - x)) (\bar{x} - x) \, dt = A(\bar{x} - x) ,$$

where

$$A = \int_0^1 F'(x + t(\bar{x} - x)) \, dt \in \mathfrak{M}$$

since D^* and $\mathfrak{M}$ are convex. Hence A is nonsingular, and from $A(\bar{x} - x) = 0$ we find $x = \bar{x}$. □

Remark: For practical applications, suitable convex sets of non-singular matrices are the sets

$$\mathfrak{M}_1 := \{A \in \mathbb{R}^{n \times n} \mid \|I - CA\| \leqq \beta\} , \tag{2}$$

where $C \in \mathbb{R}^{n \times n}$ is a fixed nonsingular matrix and $\beta \in (0, 1)$, and

$$\mathfrak{M}_2 := \{A \in \mathbb{R}^{n \times n} \mid \langle A \rangle \geqq A_0\} , \tag{3}$$

where A_0 is an M-matrix and $\langle A \rangle$ is Ostrowski's comparison matrix of A, defined by (Ostrowski [8])

$$\langle A_{ii} \rangle := |A_{ii}| , \qquad \langle A \rangle_{ik} := -|A_{ik}| \quad \text{for} \quad i \neq k .$$

Indeed, if $A \in \mathfrak{M}_1$ then by standard perturbation theory (OR2.3.2), CA and hence A is nonsingular, and

$$\|(CA)^{-1}\| \leqq (1 - \beta)^{-1} , \tag{4}$$

whereas if $A \in \mathfrak{M}_2$ then by Ostrowski [8], Satz III(8), A is nonsingular, and

$$|A^{-1}| \leqq A_0^{-1} , \tag{5}$$

where the absolute value is taken component-wise.

As a simple consequence of the semilocal inverse function theorem we obtain the well-known norm coerciveness theorem (cf. OR5.3.8).

Corollary 3: *Let D be an open, connected subset of $\mathbb{R}^n$, and let $F: D \to \mathbb{R}^n$ be a continuous function which has a nonsingular continuous derivative $F'(x)$. Suppose that F is norm coercive on D, i.e. for each $\gamma > 0$ there is a closed and bounded set $D_\gamma \subseteq D$ such that $\|F(x)\| > \gamma$ for all $x \in D \setminus D_\gamma$. Then F is a homeomorphism, i.e. for all $y \in \mathbb{R}^n$ there is a unique solution $x = F^{-1}(y) \in D$ of $F(x) = y$ depending continuously on y.*

Proof: Let E be a compact subset of $\mathbb{R}^n$. Since F is continuous, the number $\gamma := \sup \{\|F(x)\| \mid x \in E\}$ is finite. With D_γ as in the hypothesis, there is an open and bounded set D_E with $D_\gamma \subseteq D_E \subseteq D$. By construction, $F(x) \notin E$ for $x \in \partial D_E$ so that Theorem 5 applies with D_E in place of D. Hence on each compact subset E of $\mathbb{R}^n$ containing a fixed $x^0 \in \mathbb{R}^n$, the map F has a unique continuous inverse F_E^{-1} with $F_E^{-1}(F(x^0)) = x^0$. Hence F has a unique continuous inverse $F^{-1}: \mathbb{R}^n \to D$ with $F^{-1}(F(x^0)) = x^0$. Proposition 6 now guarantees uniqueness. □

We now consider applications of the semilocal inverse function theorem to the existence problem for zeros of functions.

Theorem 6: *Let D be an open and bounded subset of $\mathbb{R}^n$, and let $F: \bar{D} \to \mathbb{R}^n$ be a continuous function which has in D a nonsingular, continuous derivative. If $x^0 \in D$ and*

$$F(x) \neq sF(x^0) \quad \text{for all} \quad x \in \partial D , \qquad s \in [0, 1] \tag{6}$$

then there is a vector $x^* \in D$ with $F(x^*) = 0$. Moreover, if D is convex and $\mathfrak{M}$ is a convex set of nonsingular $n \times n$-matrices such that

$$F'(x) \in \mathfrak{M} \quad \text{for all} \quad x \in D ,$$

then x^* is the only zero of F in D.

Proof: Apply Theorem 5 with $E := \{sF(x^0) \mid s \in [0, 1]\}$, and put $x^* := F^{-1}(0)$. The uniqueness assertion follows from Proposition 7.. □

Corollary 4 (cf. Bus [1], Theorem 2.8): *Let D be an open and bounded subset of $\mathbb{R}^n$, and let $F: \bar{D} \to \mathbb{R}^n$ be a continuous function which has in D a nonsingular, continuous derivative. If $x^0 \in D$ and*

$$\|F(x)\| > \|F(x^0)\| \quad \text{for all} \quad x \in \partial D \tag{7}$$

then there is a vector $x^* \in D$ with $F(x^*) = 0$.

Proof: (7) implies (6). □

In the remainder of this section we discuss some practical ways to verify condition (6).

Proposition 8: *Let D be a convex, open, and bounded subset of $\mathbb{R}^n$, and let $F: \bar{D} \to \mathbb{R}^n$ be continuously differentiable. Suppose that $\mathfrak{M}$ is a convex set of nonsingular $n \times n$-matrices such that*

$$F'(x) \in \mathfrak{M} \quad \text{for all} \quad x \in \bar{D} .$$

If $x^0 \in D$, and if the set

$$N := \{x^0 - A^{-1}F(x^0) \mid A \in \mathfrak{M}\} \tag{8}$$

is contained in D then F has a unique zero $x^* \in D$; moreover,

$$x^* \in N. \tag{9}$$

Proof: Suppose that for some $x \in \bar{D}$ and some $s \in [0, 1]$, $F(x) = sF(x^0)$. Define $y(t) := F(x^0 + t(x - x^0))$. Then, as in the proof of Proposition 7,

$$(s - 1) F(x^0) = y(1) - y(0) = A(x - x^0)$$

for a suitable $A \in \mathfrak{M}$. Hence

$$x = sx^0 + (1 - s)(x^0 - A^{-1}F(x^0)) \tag{10}$$

is a convex combination of $x^0 \in D$ and $x^0 - A^{-1}F(x^0) \in N \subseteq D$; hence $x \in D$ since D is convex. In particular, (6) holds, and by Theorem 6, F has a unique zero $x^* \in D$. But for $x = x^*$ we have $F(x) = sF(x^0)$ with $s = 0$, whence $x = x^0 - A^{-1}F(x^0) \in N$ by (10) and (8). □

Remarks:

1. The proof also shows (take $s = 0$) that even in case that $N \nsubseteq D$, if F has a zero $x^* \in D$ then x^* is unique (by Proposition 7) and lies in N.

2. If $\mathfrak{M} = \mathfrak{M}_1$ (as in (2)) then, with

$$x^1 := x^0 - CF(x^0),$$

$x \in N$ implies

$$x - x^1 = (x^0 - A^{-1}F(x^0)) - (x^0 - CF(x^0)) = (CA)^{-1}(I - CA)(-CF(x^0)) =$$
$$= (CA)^{-1}(I - CA) = (x^1 - x^0)$$

for some $A \in \mathfrak{M}$, whence

$$\|x - x^1\| \leqq \|(CA)^{-1}\| \, \|I - CA\| \, \|x^1 - x^0\| \leqq \frac{\beta}{1 - \beta} \|x^1 - x^0\|$$

by (4). Therefore, N is a subset of

$$N_1 := \left\{x \in \mathbb{R}^n \mid \|x - x^1\| \leqq \frac{\beta}{1 - \beta} \|x^1 - x^0\|\right\},$$

and $N_1 \subseteq D$ implies the existence of a zero $x^* \in N_1$, unique in D.

3. If $\mathfrak{M} = \mathfrak{M}_2$ (as in (3)) then $x \in N$ implies that for suitable $A \in \mathfrak{M}$ we have

$$|x - x^0| = |A^{-1}F(x^0)| \leqq |A^{-1}| \, |F(x^0)| \leqq A_0^{-1}|F(x^0)|$$

by (5), whence N is a subset of

$$N_2 := \{x \in \mathbb{R}^n \mid |x - x^0| \leqq A_0^{-1}|F(x^0)|\}.$$

Again, if $N_2 \subseteq D$ then F has a zero $x^* \in N_2$, unique in D. Instead of solving a linear system to find $A_0^{-1}|F(x^0)|$, a simple upper bound can be constructed if $u, v > 0$ are known such that $A_0 u \geqq v$. Then there is a constant $\alpha > 0$ such that $|F(x^0)| \leqq \alpha u$, and since $A_0^{-1} \geqq 0$, this implies that $A_0^{-1}|F(x^0)| \leqq \alpha A_0^{-1} u \leqq \alpha v$.

4. If $\mathfrak{M} = \{A \in \mathbb{R}^{n \times n} \mid \underline{A} \leqq A \leqq \bar{A}\}$ is a matrix interval then interval arithmetic (cf. MOORE [4]), can be used in various ways to obtain vector intervals containing N. The resulting bounding methods for zeros of equations are often called interval Newton methods. The first existence test for an interval Newton method, a special case of the above proposition, was proved by NICKEL [6] for the special case that the enclosure of N is computed as

$$N_3 := x^0 - \mathfrak{B}F(x^0),$$

where $\mathfrak{B}$ is a matrix interval containing all A^{-1} with $A \in \mathfrak{M}$. NICKEL's proof, however, is less elementary in that it uses Brouwer's fixpoint theorem. A number of other interval Newton methods have been proposed in the last few years; see e.g. KRAWCZYK [3] and the references there.

References

1 BUS, J. C. P., Numerical Solution of Systems of Nonlinear Equations, Math. Centre Tracts 122, Math. Centrum, Amsterdam 1980.
2 DIEUDONNÉ, J., Foundations of Modern Analysis, Enlarged and corrected printing, Academic Press, New York-London 1969.
3 KRAWCZYK, R., Intervalliterationsverfahren, Bericht Math.- Stat. Sect. Forschungszentrum Graz, 186, 1982.
4 MOORE, R. E., Methods and Applications of Interval Analysis, SIAM Publications, Philadelphia 1979.
5 NEUMAIER, A., Simple bounds for zeros of systems of equations, in: Iterative Solution of Nonlinear Systems of Equations, Lecture Notes in Math. 953, Springer-Verlag, Berlin-Heidelberg-New York 1982, pp. 88–105.
6 NICKEL, K., On the Newton method in Interval Analysis, MRC Tech. Sum. Rep. (Univ. of Wisconsin, Madison) 1136, 1971.
7 ORTEGA, J. M.; RHEINBOLDT, W. C., Iterative Solution of Nonlinear Equations in Several Variables, Academic Press, New York-London 1970.
8 OSTROWSKI, A. M., Über die Determinanten mit überwiegender Hauptdiagonale, Comment. Math. Helvet. 10 (1937), 69–96.
9 POURCIAU, B. H., Homeomorphisms and generalized derivatives, J. Math. Anal. Appl. 93 (1983), 338–343.
10 SCHWETLICK, H., Numerische Lösung nichtlinearer Gleichungen, VEB Deutscher Verlag der Wiss., Berlin 1978.

Received July 7, 1983, revised version December 1, 1983

Address: Dr. ARNOLD NEUMAIER, Institut für Angewandte Mathematik der Albert-Ludwigs-Universität, Hermann-Herder-Straße 10, D-7800 Freiburg i. Br., BRD

BOOK REVIEWS

Robinson, A. R. (ed.), Eddies in Marine Science. Berlin-Heidelberg-New York-Tokyo, Springer-Verlag 1983. XXV, 609 S., 268 Abb., DM 120,–. US $ 46.60. ISBN 3-540-12253-2

Unter Wirbeln wird in dieser Monographie eine große Klasse räumlich und zeitlich veränderlicher Meeresströmungen verstanden; die Variabilitätsbereiche erstrecken sich dabei von 10 bis zu einigen hundert Kilometern bzw. über Wochen und Monate. Die Wirbelwegung wird ausgelöst durch Corioliskraft und Gravitation und verläuft fast horizontal. Um ihre zeitliche Entwicklung und räumliche Dynamik zu verstehen, muß eine kleine vertikale Geschwindigkeitskomponente und eine geringe Abweichung von der rein geostrophischen Impulsbilanz berücksichtigt werden. Solche Wirbelströme gibt es in allen Weltmeeren, besonders ausgeprägt sind sie in der Umgebung intensiver Meeresströmungen wie z. B. des Golfstromes.

Man ist heute in der Lage, diese Erscheinungen, ähnlich wie in der Atmosphäre, großräumig zu beobachten und Modellvorstellungen zu entwickeln. Weltweite nationale und internationale ozeanographische Forschungsunternehmnungen werden nicht zuletzt wegen ihrer wachsenden Bedeutung für die Fischerei, die Mineral- und Energiegewinnung und das Transportwesen durchgeführt. Die mit den Wirbeln verbundenen Temperaturschwankungen beeinflussen die akustischen Eigenschaften der ozeanischen Wassermassen.

Die Monographie bringt in ihrem ersten Teil eine nach regionalen Gesichtspunkten gegliederte Beschreibung und Analyse wirbelartiger Meeresströmungen. Eigene Kapitel sind dem Golfstrom, dem westlichen Nordatlantik, dem Nordostatlantik, dem Nordpazifik, dem nördlichen Eismeer, dem tropischen Äquatorialgebiet, dem Indischen Ozean, dem Südpazifik einschließlich des ostausstralischen Stroms und dem Südpolarstrom gewidmet. Dabei sind einige Meeresteile wie z. B. der westliche Nordatlantik sehr viel ausführlicher behandelt als andere.

Seit den siebziger Jahren werden in Verbindung mit großräumigen Forschungsprogrammen numerische Modelle für die Zirkulation im Ozean entwickelt, um die Wirkung von Wirbeln mittlerer Größenordnung zu studieren. Periodische und regionale Modelle dienen dazu, das Zusammenwirken mittel- und großräumiger Zirkulationen im Ozean zu untersuchen. Mit diesen Fragen, einschließlich der zweidimensionalen Turbulenz beschäftigt sich der zweite Teil. Der dritte Teil befaßt sich mit dem Einfluß der Wirbel auf klimatische Erscheinungen, auf Küstenregionen und auf biologische Prozesse. Ein eigener Abschnitt ist dem Einfluß auf die Schallausbreitung gewidmet. Das Schlußkapitel hat die neuere ozeanographische Meßtechnik zum Gegenstand.

An der Monographie waren neben dem Herausgeber mehr als 30 namhafte Autoren beteiligt. Das Literaturverzeichnis enthält etwa 850 Titel.

Berlin J. Förste

Duff, M. J. / Isham, C. J. (eds.), Quantum Structure of Space and Time. Proceedings, London 1981. Cambridge et al., Cambridge University Press 1982. XIII, 432 S., £ 25.00. ISBN 0 521 24732 2

Der vorliegende Band faßt unter der im Titel formulierten Aufgabenstellung Untersuchungsergebnisse zusammen, die während einer dreiwöchigen Tagung (je eine Woche für die Teile I, II und III) zur Quantisierung des Gravitationsfeldes und seiner Vereinigung mit den anderen Fundamentalkräften der Natur diskutiert wurden.

Teil I konzentriert sich in Fortsetzung der Grundgedanken von Einsteins Allgemeiner Relativitätstheorie auf geometrisch-topologische Aspekte einer Quantengravitation: kanonische Quantisierung, kovariante Pathintegrale, Regge-Kalkül, spontane Symmetriebrechung, Instantons usw. Der mit „Supergravität" überschriebene Teil II greift die offenen Probleme auf, die Lagrangians und ihre Hilfsfelder, supersymmetrische Yang-Hills-Theorien und die oben genannte Superunifikation zu finden, unter methodischer Einbeziehung von kompakten Extradimensionen (verallgemeinerte Kaluza-Klein-Theorien), Regge-Pols, Schwarze-Löche-Solitonen usw. Der letzte und für den außenstehenden Leser vielleicht interessanteste Teil III zieht die Schlußfolgerungen für unser Verständnis einer Kosmologie unter expliziter Einbeziehung der möglichen Zustände des frühen Universums. Er berührt die Fragen nach dem Grund für das Verschwinden der kosmologischen Konstante und der damit verbundenen „Feinabstimmung" der freien Parameter des Universums, Fragen der Erhaltungsgrößen und von Horizonteffekten, Phasenübergängen im Zusammenhang mit der großräumigen Homogenität und Isotropie des Kosmos, der für die Galaxienbildung zur Verfügung stehenden Zeitskala, der Baryonenasymmetrie und ihrer spezifischen Entropie, bis hin zu einer Neuformulierung des „Anthropischen Prinzips". Das unter wesentlicher Mitwirkung führender Vertreter auf diesem Gebiet entstandene Werk kann als eine gute Übersicht empfohlen werden.

Berlin G. Wallis

Nečas, J., Introduction to the Theory of Nonlinear Elliptic Equations. Leipzig, BSB B.G. Teubner VG 1983. 202 S., 3 Abb., M 19,–. BN 6660493 (Teubner-Texte zur Mathematik 52)

In dem Buch werden die quasilinearen elliptischen Systeme partieller Differentialgleichungen zweiter Ordnung in Divergenzform und ihre Randwertaufgaben behandelt. Der Autor wählt den funktionalanalytischen Zugang, dem zufolge das System von m Differentialgleichungen für m gesuchte Funktionen und die Randbedingungen einen stetigen Operator definieren, der einen reflexiven Banachraum — Teilraum eines Sobolevraumes — in seinen dualen Raum abbildet. Er hat vor allem Anwendungen in Physik und Technik vor Augen, bei denen der Operator die (Material-)Eigenschaften eines modellierten Systems beschreibt und die rechte Seite die äußeren Kräfte repräsentiert.

Nach einer knappen Einführung und nach der Bereitstellung der Sobolev- und Morrey-Campanato-Räume mit einigen ihrer Eigenschaften werden in Kapitel 3 die Randwertaufgaben formuliert und Existenz- und Eindeutigkeitsfragen behandelt. Es wird eine umfangreiche Klasse von Randbedingungen in Gebieten mit Lipschitz stetigem Rand betrachtet. Den Existenz- und Eindeutigkeitsaussagen liegen die funktionalanalytischen Konzepte der Variationsrechnung und der monotonen Operatoren zugrunde. Die Kapitel 5 und 6 über die Regularität von Lösungen bilden den Kern des Buches. Sie spiegeln die neuere Entwicklung auf diesem Gebiet wider, an der der Verfasser des Buches selbst hervorragenden Anteil hat. Im Kapitel 5 werden die lokale quadratische Integrierbarkeit der zweiten Ableitungen der Lösungen und ihre quadratische Integrierbarkeit bis an den Rand untersucht. Unter recht speziellen und etwas unnatürlichen Voraussetzungen werden auch Abschätzungen für den Betrag des Gradienten angegeben. Im Kapitel 6 wird die Hölderstetigkeit der ersten Ableitungen untersucht, die für Systeme im allgemeinen nicht garantiert ist. Im Innern wird sie unter zusätzlichen Voraussetzungen durch eine vom Autor zusammen mit Giaquinta entdeckte sog. Liouvillebedingung gesichert, aber ohne diese Bedingung besteht sie nur außerhalb einer gewissen Ausnahmemenge.

Der Referent zählt zu den besonderen Vorzügen des Buches, daß es nahezu ausschließlich Systemen von Differentialgleichungen gewidmet ist, die in anderen Monographien häufig nur als Ergänzung zum skalaren Fall skizziert werden, daß der Leser schnell zu den wesentlichen Aussagen vordringt und auf eine Fülle von Gesichtspunkten hingewiesen wird, die man bei dem handlichen Buch nicht vermuten würde, und daß diese Anregungen durch sorgfältige Literaturhinweise ergänzt werden. Das Buch ist auch insofern sehr anregend, als Beweise oft nur skizziert sind und in den Sätzen nicht immer alle Voraussetzungen angegeben werden. Das vierte Kapitel über Näherungsverfahren fällt etwas aus dem Rahmen des Buches, da die funktionalanalytischen Sätze dort nicht auf Gleichungssysteme angewendet werden wie etwa im Kapitel 3.

Berlin G. Albinus

SHORT COMMUNICATIONS

ZAMM · Z. Angew. Math. u. Mech. 65 (1985) 1, 57–59

ATANACKOVIC, T. M.[1]; ACHENBACH, M.

A Least Square Solution and Error Estimate for some Vibration Problems

An error estimate procedure, based on the value of the mean square residual, suitable for a class of initial value problems is proposed. The method is illustrated by two examples from the non-linear vibration theory.

1. Introduction

The least-squares method represents an attractive procedure for finding approximate solutions of non-linear problems in physics and engineering sciences. The major advantages of the least-squares method are its simplicity and the fact that it could be used without any restrictions, for both, boundary and initial value problems.

Further advantage is that the method is used directly, i.e. it is applied to given differential equations. This is an important property, when compared with variational methods, for example, in which one has first to construct the Lagrangian for given equations, and then to apply the so-called direct methods of variational calculus (such as Ritz and Kantorovich procedures). On the other hand, when treated as a variational principle [1], [5] (in which case it is known as Gauss principle of least constraint) the least square method has the important property of being the minimum principle. For comparison of least squares and other methods for approximate solution of engineering problems see, for example, [1].

The mean square residual, which is minimized in the least squares method, has also theoretical significance since error bounds can be derived in terms of it. Thus minimization of the mean square residual leads to the best possible bounds for the error.

There are many error estimating procedures based on the value of the mean square ressidual [1], [2], [3], [4]. Each is applicable to a certain class of problems. Our intention in this note is to present a method of estimating the error of an approximate solution via the mean square residual, that is suitable for a class of initial value problems frequently occurring in vibration theory. The procedure that we propose here is different to those given in [4] and [5]. In the present analysis the integral inequalities obtained in [6], [7] and [8] will be of central importance. Finally we shall demonstrate the method on two concrete problems.

2. Error estimating procedure

Let us consider the following non-linear second order differential equation

$$\ddot{q} - F(q, t) = 0\,, \qquad t > 0\,, \tag{1}$$

where $(\dot{\cdot}) = \frac{\mathrm{d}}{\mathrm{d}t}(\ )$. We assume that together with (1) the following initial conditions are given

$$q(0) = A\,, \qquad \dot{q}(0) = B\,. \tag{2}$$

In the analysis that follows we shall assume that the problem (1), (2) possesses a unique $C^2[0, T]$ solution. We denote this (exact) solution by q.

Let $Q(t)$ be an admissible trial function, i.e. a $C^2[0, T]$ function satisfying the initial conditions (2). The mean square residual over an arbitrary time interval $[0, T]$, associated with Q is given by the following functional

$$E^T(Q) = \int_0^T R^2\,\mathrm{d}t\,, \tag{3}$$

where $R \equiv \ddot{Q} - F(Q, t)$ is the residual. From (3) and (1) we have that $E(q) = 0$. Let $\delta q = Q - q$. Then, the error δq satisfies the following initial conditions

$$\delta q(0) = 0; \qquad \delta\dot{q}(0) = 0\,. \tag{4}$$

Using the mean-value theorem for functionals, we may write (3) as

$$E^T(Q) = E^T(q) + \delta E^T(q, \delta q) + \tfrac{1}{2}\,\delta^2 E^T(\psi, \delta q) \tag{5}$$

where δE^T and $\delta^2 E^T$ denote the first and second variation, respectively, and where

$$\psi = q + \eta\delta q\,, \qquad 0 \leqq \eta \leqq 1\,. \tag{6}$$

It is easy to see that the first variation of E^T at q is equal to zero, i.e.

$$\delta E^T(q, \delta q) = 0\,. \tag{7}$$

Therefore (5) becomes

$$E^T(Q) = \tfrac{1}{2}\,\delta^2 E^T(\psi, \delta q)\,. \tag{8}$$

Calculating the second variation of (3) and using it in (8) we have

$$E^T(Q) = \int_0^T [\delta\ddot{q}^2 + A(\psi)\,\delta q\delta\ddot{q} + B(\psi)\,\delta q^2]\,\mathrm{d}t\,, \tag{9}$$

where

$$A(\psi) = -2\left(\frac{\partial F}{\partial Q}\right)_\psi, \qquad B(\psi) = \left[\left(\frac{\partial F}{\partial Q}\right)_\psi\right]^2 - \left(\frac{\partial^2 F}{\partial Q^2}\right)_\psi R(\psi)\,. \tag{10}$$

In the further analysis we shall distinguish the following two cases:

Case i): Suppose that $B(\psi) > 0$ and there exist two constants (depending on Q) such that

$$|A(\psi)| \leqq C_1; \qquad B(\psi) \geqq C_2 \geqq 0\,, \qquad t \in (0, T)\,, \tag{11); (12}$$

then by using (11) and (12) in (9) we have

$$E^T(Q) \geqq \int_0^T [\delta\ddot{q}^2 - C_1\delta q\delta\ddot{q} + C_2\delta q^2]\,\mathrm{d}t\,. \tag{13}$$

Before proceeding further, we make a few remarks concerning the conditions (11) and (12). In general, it is hard, or impossible, to find estimates of the form (11) and (12). However, in vibration problems, we usually have either energy integral or some other information (boundness for example) about the exact solution q. Those additional properties of q may help in obtaining (11) and (12). For example, with the help of the energy integral (which is readily obtainable for autonomous systems) estimates (11) and (12) could be easily derived.

To transform (13) we first use the following inequality due to DAS [7]

$$\int_0^T |\delta q\delta\ddot{q}|\,\mathrm{d}t + \leqq \tfrac{1}{2}\,(\tfrac{2}{3})^{1/2}\,T^2 \int_0^T \delta\ddot{q}^2\,\mathrm{d}t\,, \tag{14}$$

which is applicable because of (4). Using (14), inequality (13) becomes

$$E^T(Q) \geqq ||\delta\ddot{q}||_{L_2}[1 - T^2C] + C_2||\delta q||_{L_2}^2 \tag{15}$$

where $C = C_1\,\frac{1}{4}\,(2/3)^{1/2}$ and where

$$||\delta q||_{L_2} = \left(\int_0^T \delta q^2\,\mathrm{d}t\right)^{1/2} \tag{16}$$

is the L_2 norm. To obtain a relation between the L_2 norm of δq and $\delta\ddot{q}$ we use the inequality of TROESCH [8],

$$\int_0^1 h(t)\,\dot{f}^2\,\mathrm{d}t \geqq \frac{\pi^2}{4}\int_0^1 h\,\mathrm{d}t \int_0^1 f^2\,\mathrm{d}t\,, \tag{17}$$

with $h(t) = 1$. Substituting first $f = \delta q$ and then $f = \delta\dot{q}$ we get

$$||\delta\ddot{q}||_{L_2} \geqq \frac{\pi}{2T}\,||\delta\dot{q}||_{L_2}; \qquad ||\delta\dot{q}||_{L_2} \geqq \frac{\pi}{2T}\,||\delta q||_{L_2}\,. \tag{18}$$

From (18) and (15) we have the following estimate

$$||\delta q||_{L_2} \leqq \left\{\frac{E^T(Q)}{C_2 + \left(\frac{\pi}{2T}\right)^4 [1 - T^2C]}\right\}^{1/2}. \tag{19}$$

[1]) On leave from the University of Novi Sad, 21000 Novi Sad, Yugoslavia.

In writing (19) we assumed that $1 - T^2C \geqq 0$. Therefore, in each particular case we will choose the time interval in which we estimate the error so that this condition is satisfied.

To estimate the L_∞ norm,

$$||\delta q||_{L_\infty} = \sup_{t\in(0,T)} |\delta q| , \tag{20}$$

we shall first use the following inequality [6 p. 147]

$$||\delta \dot q||^2_{L_2} \leqq \varepsilon_1 ||\delta \ddot q||_{L_2} + K(\varepsilon_1) ||\delta q||^2_{L_2} , \tag{21}$$

where $\varepsilon_1 > 0$ and $K(\varepsilon_1) = (1/\varepsilon_1) + 12/T^2$. From (21) we have

$$||\delta q||^2_{L_2} \geqq \frac{1}{K(\varepsilon_1)} \{||\delta \dot q||^2_{L_2} - \varepsilon_1 ||\delta \ddot q||^2_{L_2}\} . \tag{22}$$

Using $(18)_1$ and (22) in (15) we get

$$E^T(Q) \geqq \left\{\left[(1 - T^2C) - \frac{\varepsilon_1}{K(\varepsilon_1)} C_2\right] \left(\frac{\pi}{2T}\right)^2 + \frac{C_2}{K(\varepsilon_1)}\right\} ||\delta \dot q||^2_{L_2} . \tag{23}$$

However, Cauchy's inequality and (4) imply

$$||\delta q||_{L_\infty} \leqq T^{1/2} ||\delta \dot q||_{L_2} , \tag{24}$$

so that (23) gives

$$||\delta q||_{L_\infty} \leqq \left\{T \frac{E^T(Q)}{\left[(1 - T^2C) - \frac{\varepsilon_1}{K(\varepsilon_1)} C_2\right]\left(\frac{\pi}{2T}\right)^2 + \frac{C_2}{K(\varepsilon_1)}}\right\}^{1/2} , \tag{25}$$

Case ii): Suppose that $B(\psi) \leqq 0$ and there exist two constants (depending on Q) such that

$$|A(\psi)| \leqq \hat C_1 , \qquad |B(\psi)| \leqq \hat C_2 . \tag{26), (27}$$

Using a similar procedure as before, we get

$$||\delta q||_{L_2} \leqq \left\{\frac{E^T(Q)}{-\hat C_2 + \left(\frac{\pi}{2T}\right)^4 [1 - T^2 \hat C]}\right\}^{1/2} , \tag{28}$$

where $\hat C = \hat C_1 \frac{1}{4} (2/3)^{1/2}$. Also, instead of (25) we have

$$||\delta q||_{L_\infty} \leqq \left\{T \frac{E^T(Q)}{\left(\frac{\pi}{2T}\right)^2 [1 - T^2 \hat C] - \hat C_2 \left(\frac{2T}{\pi}\right)^2}\right\}^{1/2} . \tag{29}$$

In writing (29) we used (13), (18), (26) and (27).

Remark 1: Since (21) is valid for any $\varepsilon_1 > 0$ we may choose it in the optimal way, i.e. so as to make the bound given by (25) minimal.

Remark 2: In certain cases it is not possible to estimate $A(\psi)$ and $B(\psi)$ by constants depending on Q only, but with a constant depending on Q and $||\delta q||_{L_\infty}$. Our error estimate procedure in this case could be modified to follow the analysis of [9].

3. Examples

In this section we shall apply the method of estimating error, developed in the previous part to two concrete problems.

A) As a first example let us consider the Duffing equation

$$\ddot q + q + \varepsilon q^3 = 0 , \qquad \varepsilon > 0 , \tag{30}$$

with the following initial conditions

$$q(0) = A , \qquad \dot q(0) = 0 . \tag{31}$$

We assume an approximate solution to (30), (31) in the form

$$Q = A \cos \omega t \tag{32}$$

and chose $T_1 = \pi/2\omega$. The mean square residual then becomes

$$E^{T_1} = \int_0^{\pi/2\omega} [(-A\omega^2 + A) \cos \omega t + \varepsilon A^3 \cos^3 \omega t]^2 \, dt$$

$$= \frac{1}{\omega} \left[A^2(1 - \omega^2)^2 \frac{\pi}{4} + \frac{3\pi}{8} \varepsilon A^4(1 - \omega^2) + \varepsilon^2 A^6 \frac{5\pi}{32}\right], \tag{33}$$

Minimizing (33) with respect to ω (i.e. determining ω from the condition $dE^{T_1}/d\omega = 0$) we get

$$\bar\omega^2 = \tfrac{1}{3} + \tfrac{2}{3} \{1 + \tfrac{3}{2} \varepsilon A^2 + \tfrac{39}{16} \varepsilon^2 A^4 + \tfrac{1}{4} \varepsilon A^2\}^{1/2} . \tag{34}$$

We note that for $\varepsilon \ll 1$, (34) gives

$$\hat\omega^2 = \bar\omega^2 \cong 1 + \tfrac{3}{4} \varepsilon A^2 , \tag{35}$$

in agreement with the results obtained by other methods [10].

To estimate the error, we first note that the energy integral

$$\dot q^2 + q^2 + \varepsilon \frac{q^4}{2} = \left(1 + \frac{\varepsilon A^2}{2}\right) A^2 \tag{36}$$

with $\varepsilon > 0$ imply that $q(t) \leqq A$. Therefore we have

$$\hat C_1 = 2(1 + 3\varepsilon A^2); \qquad \hat C = \frac{\sqrt{6}}{6} (1 + 3\varepsilon A^2);$$

$$\hat C_2 = 6\varepsilon A^2(1 + \tfrac{3}{2} \varepsilon A^2) . \tag{37}$$

From (37) and (34) we see that the condition $1 - T_1^2 \hat C > 0$ is not satisfied for $T_1 = 2\pi/\bar\omega$. Therefore we can estimate the error in a time interval that is smaller than the half period of the approximate solution (32) with $\omega = \bar\omega$. Thus we take $T = \pi/4\bar\omega$. The value of the mean square residual becomes

$$E^T = \frac{1}{\bar\omega} \left[A^2(1 - \bar\omega^2)^2 \frac{\pi + 2}{8} + 2\varepsilon A^4(1 - \bar\omega^2) \frac{3\pi + 8}{32} + \right.$$

$$\left. + \varepsilon^2 A^6 \frac{15\pi + 44}{192}\right] . \tag{38}$$

The condition $1 - \hat C T^2 > 0$ is now satisfied, so that (28) gives

$$||\delta q||_{L_2} \leqq$$

$$\left\{\frac{E^T(\bar\omega)}{4\bar\omega^2 \left[1 - \frac{\pi^2}{4\bar\omega^2}\left[\frac{\sqrt{6}}{6} (1 + 3\varepsilon A^2)\right]\right] - 6\varepsilon A^2(1 + \frac{3}{2} \varepsilon A^2)}\right\}^{1/2} \tag{39}$$

where $\bar\omega$ and $E^T(\bar\omega)$ are given by (34) and (38), respectively.

B) As a second example we consider the following initial value problem

$$\ddot q + \omega_0^2 q = \alpha + \varepsilon q^2 , \tag{40}$$

with

$$q(0) = A , \qquad \dot q(0) = 0 . \tag{41}$$

The problem (40), (41) arises in many branches of mathematical physics and has been recently treated in [11] (see also the discussion in the same Journal following that paper).

In the present case we take

$$Q = (A - \alpha) \cos \omega t + \alpha . \tag{42}$$

Again we choose $T_1 = \pi/2\omega$ so that (3) becomes

$$E^{T_1} = \frac{1}{\omega} \left\{(\omega_0^2 - \omega^2)^2 (A - \alpha)^2 \frac{\pi}{4} - 2(A - \alpha) (\omega_0^2 - \omega^2) \times \right.$$

$$\times \varepsilon \left[\frac{2}{3} (A - \alpha)^2 + \frac{\pi}{2} (A - \alpha) \alpha + \alpha^2\right] +$$

$$+ \varepsilon^2 \left[(A - \alpha)^4 \frac{3\pi}{16} + \frac{8}{3} (A - \alpha)^3 \alpha + \frac{3}{2} \pi (A - \alpha)^2 \alpha^2 + \right.$$

$$\left. \left. + 4(A - \alpha) \alpha^3 + \alpha^4 \frac{\pi}{2}\right]\right\} . \tag{43}$$

Further the procedure requires to solve a biquadratic equation to determine ω. To simplify the procedure we consider the special case

$$\omega_0^2 = 0 , \qquad \alpha = 0 , \qquad -\varepsilon = k > 0 . \tag{44}$$

Then (40) corresponds to a nonlinear oscillator treated in [12]. The expression (43) then becomes

$$E^{T_1} = \frac{1}{\omega} \left[k^2 A^4 \frac{3\pi}{16} + \omega^4 A^2 \frac{\pi}{4} - \frac{4}{3} A^3 k \omega^2\right] . \tag{45}$$

The condition $dE^{T_1}/d\omega = 0$ leads to

$$\bar\omega = 0.925984 \, A^{1/2} k^{1/2} . \tag{46}$$

The value for $\bar\omega$ agrees well with the value $\omega = 0.921317 A^{1/2} k^{1/2}$ obtained in [12] by the methods of harmonic balance. Again we take $[0, T]$, $T = \pi/4\bar\omega$, as a time interval in which we shall estimate the error of the approximate solution (42), (46). With the following values for the constants in (26) and (27)

$$\hat C_1 = 4kA , \qquad \hat C = \sqrt{\tfrac{2}{3}} kA , \qquad \hat C_2 = 2k^2 A^2 , \tag{47}$$

inequality (29) gives

$$||\delta q||_{L_\infty} \leqq 0.08482235A\,. \tag{48}$$

4. Conclusions

In this paper we proposed a procedure for estimating the error of an approximate solution by the use of the mean square residual. The method is suitable for initial valve problems and is based on inequalities given in [6], [7] and [8]. In applying the procedure one has to distinguish between the time interval in which the mean square residual is minimized to obtain an approximate solution (in our notation $[0, T_1]$) and the time interval in which the error is estimated (in our notation $[0, T]$). The time interval $[0, T]$ has to be choosen so that the restrictions imposed by the method are satisfied.

References

1 FINLAYSON, B. A., The Method of Weighted Residuals and Variational Principles, Academic Press New York 1972.
2 CIARLET, P. G.; SCHULTZ, M. H.; VARGA, R. S., Numerical methods of high-order. Accuracy for nonlinear boundary value problems, Numerische Mathematik 9, 344–430 (1967).
3 SIGILLITO, V. G., Pointwise bounds for solutions of parabolic equations, SIAM Rev. 9, 581–585 (1967).
4 DJUKIC, DJ. S.; ATANACKOVIC, T. M., Contribution to error estimate, J. Math. Anal. Appl. 88, 183–195 (1982).
5 DJUKIC, DJ. S.; ATANACKOVIC, T. M., The least squares method: Kantorovich; approach, Int. J. Heat Mass Transfer 24, 443–448 (1981).
6 MITRINOVIĆ, D. S., Analytic Inequalities, Springer Verlag, Berlin 1970, p. 147.
7 DAS, K. M., An inequality similar to Opial's inequality, Proc. Amer. Math. Soc. 22, 258–261 (1969).
8 TROESCH, B. A., Integral inequalities for two functions, Arch. Rational Mech. Anal. 24, 128–140 (1967).
9 DJUKIC, DJ. S.; ATANACKOVIC, T. M., An extremum variational principle and error estimate procedure for $q^{IV} = f(q, x)$, Math. Proc. Cambridge Philos. Soc. 92, 307–316 (1982).
10 NAYFEH, A. H., Perturbation Methods, John Wiley, New York 1973.
11 MICKENS, R. E., A uniformly valid asymptotic solution for $y'' + y = \alpha + \varepsilon y^2$, J. Sound Vibration 76, 150–152 (1981).
12 NAYFEH, A. H.; MOOK, D. T., Nonlinear Oscillations, John Wiley, New York 1979.

Received February 28, 1983

Address: Dr. T. M. ATANACKOVIC; Dr. M. ACHENBACH, Fachbereich 9, Hermann-Föttinger Institut, TU Berlin, Berlin-West 12

ZAMM · Z. Angew. Math. u. Mech. 65 (1985) 1, 59–61

TOMSKI, L.

Prebuckling Behaviour of Compound Column-Direct Nonlinear Analysis

The prebuckling behaviour of the centrically loaded geometrically perfect compound column with pinned ends is analysed. The column is constructed from the coaxial prismatic bar of different axial and flexural rigidities. The bars are rigidly joined together at their ends. The equations of equilibrium include the nonlinear strain term due to the bending of the bars.

Problem formulation

Static stability analysis of a compound column based on the direct nonlinear theory is performed. The equations of equilibrium of the column in the lateral and axial directions include nonlinear strain terms due to bending of the bars (see e.g. [1–3]).

The solutions for an eccentrically quasi-statically loaded two-bar frame (cf. [4]) and for a geometrically imperfect strut of a linear-elastic anisotropic material under axial impact (cf. [3]) yield buckling behaviours which correlate well with the experimental results (cf. [5] and [3], respectively).

We consider the buckling of a geometrically perfect compound bar constructed from coaxial prismatic members of different materials with Young's moduli E_1 and E_2 (Fig. 1). By A_1I_1 and A_2I_2 are denoted the cross-sectional area and the second moment of the area of inner and exterior members, respectively. The column is pinned at its ends but both bars are rigidly joined together at their ends. The equivalent buckling length is l.

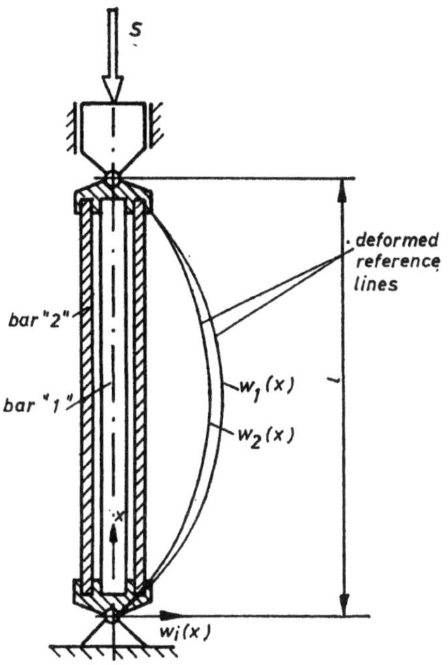

Fig. 1. Geometry of the compound bar

If $w_i \equiv w_i(x)$, $M_i \equiv M_i(x)$ and S_i denote the lateral deflection, bending moment and axial force of the i-th bar, respectively, the equations of equilibrium [3] are

$$(M_{i,x} + S_i w_{i,x})_{,x} = 0\,, \tag{1}$$

$$S_{i,x} = 0\,. \tag{2}$$

For the linear Hooke-type material and with the axial deformation (ε_i)-displacement nonlinear relation

$$\varepsilon_i = u_{i,x} + \tfrac{1}{2}\, w_{i,x}^2 \tag{3}$$

the equations of equilibrium (1) and (2) become

$$E_iI_iw_{i,xxxx} + S_iw_{i,xx} = 0\,, \tag{4}$$

$$(u_{i,x} + \tfrac{1}{2}\, w_{i,x}^2)_{,x} = 0\,, \tag{5}$$

where u_i is the axial displacement.

The geometric boundary and continuity conditions are given by:

$$w_1(0) = w_2(0) = w_1(l) = w_2(l) = 0; \tag{6_{1-4}}$$

$$w_{1,x}(0) = w_{2,x}(0); \qquad w_{1,x}(l) = w_{2,x}(l); \tag{7_{1-2}}$$

$$u_1(0) = u_2(0) = 0; \qquad u_1(l) = u_2(l); \tag{8_{1-3}}$$

$$E_1I_1w_{1,xx}(0) + E_2I_2w_{2,xx}(0) = 0; \tag{9_1}$$

$$E_1I_1w_{1,xx}(l) + E_2I_2w_{2,xx}(l) = 0\,. \tag{9_2}$$

Solving the problem

Denoting $M_0 = E_1I_1w_{1,xx}(0)$, $M_1 = E_1I_1w_{1,xx}(l)$ and integrating equation (4) by virtue of the condition (6_{1-4}) and if (9_{1-2}) are used, then the well-known general solution for lateral deflection is reduced to

$$w_i(x) = (-1)^i \left[\frac{M_0}{S_i} + \frac{M_1 - M_0}{S_i l}\, x - \frac{1}{S_i \sin k_i l} \times \right.$$
$$\left. \times\, (M_1 - M_0 \cos k_i l) \sin k_i x + - \frac{M_0}{S_i} \cos k_i x\right], \tag{10}$$

where $k_i^2 = S_i/E_iI_i$.

Taking into account the two remaining boundary conditions (7_{1-2}) to which the solution (10) is subjected, we obtain a transcendental equation with respect to the internal forces S_1 and S_2 in the form:

$$F(S_1, S_2) = \frac{k_1}{S_1}\, \mathrm{tg}\, \frac{k_1 l}{2} + \frac{k_2}{S_2}\, \mathrm{tg}\, \frac{k_2 l}{2} = 0\,. \tag{11}$$

Moreover, on the basis of the conditions (7_{1-2}) the relation $M_1 = M_0$ is obtained. Denoting $M = M_0 = M_1$ the solution (10) can be written as

$$w_i(x) = (-1)^i \frac{M}{S_i} \left[1 - (1 - \cos k_i l) \frac{\sin k_i x}{\sin k_i l} - \cos k_i x\right]. \tag{12}$$

Inserting the solution (12) into the equation (5) and after integrating the equation in conjunction with (8_{1-2}) we finally have

$$u_i(x) = -\frac{S_i}{A_i E_i} x - M^2 \frac{k_i}{2S_i^2} \frac{1 - \cos k_i l}{\sin^2 k_i l} \times$$
$$\times [k_i x - \sin k_i x \cos k_i (l - x)] \tag{13}$$

Substituting the solution (13) into the condition (8_3) we obtain the bending moment in the following form:

$$M = \left[\frac{-2l\left(\frac{S_2}{A_2E_2} - \frac{S_1}{A_1E_1}\right)}{\frac{k_2}{S_2^2}\left(\frac{k_2 l}{\sin k_2 l} - 1\right) \operatorname{tg}\frac{k_2 l}{2} - \frac{k_1}{S_1^2}\left(\frac{k_1 l}{\sin k_1 l} - 1\right) \operatorname{tg}\frac{k_1 l}{2}}\right]^{1/2}. \tag{14}$$

Behaviour of the system

At the increase of the external force S, if there is no bending, we have the relationship of internal forces in the bars as the following

$$\frac{S_2}{S_c} = \frac{A_2E_2}{A_1E_1} \frac{S_1}{S_c}. \tag{15}$$

This linear relationship is denoted by "1" in Fig. 2. The values of forces S_1 and S_2 are shown as their ratio to the critical load which value corresponds to the point c and is equal to

$$S_c = \frac{\pi^2(E_1I_1 + E_2I_2)}{l^2}. \tag{16}$$

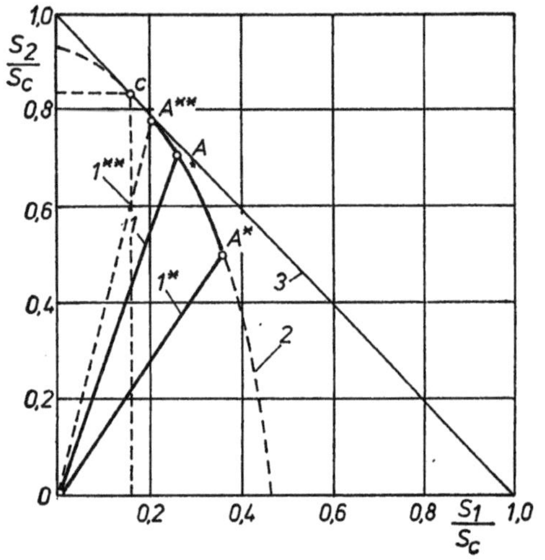

Fig. 2. Relationship between internal forces S_1 and S_2 within the range of the quasi-statical increase of external load, from zero to the buckling value

If the total load S reaches the point A (Fig. 2) the further change of internal forces S_1 and S_2 in particular bars will occur according to the curve defined by equation (11) (curve "2", Fig. 2). Then at the point A there appears the bending moment in the system; described by the relationship [14]. At further increase of the total external load S can reach the point c. This value of the force produces the buckling of the system according to the linear theory of stability. This means that the value of force S_c will be equal to the buckling load which can be obtained if it is assumed that the centres of gravity of the cross sections of the components coincide.

At the given values of axial rigidity A_2E_2 and flexural stiffnesses E_1I_1 and E_2I_2, the axial rigidity of the bar "1" can be within the range

$$E_1A_1^{**} \leq E_1A_1 \leq E_1A_1^{*}. \tag{17}$$

For the chosen design of the column which concerns the fact that the bar "1" is inside the bar "2" the maximal cross-sectional area A^* is obtained when the internal diameter of the bar "1" is equal to zero. The minimal area A^{**} is obtained when the external diameter of that bar will be equal to the internal diameter of the bar "2".

The increase of the forces in particular bars, when bending does not occur yet, at axial rigidities E_1A_1 and $E_1A_1^*$, is shown in Fig. 2 by lines "1" and "1*", respectively. The further increase of the external load causes the values of S_1, and S_2 forces to change according to the curve "2" (from points A or A^* to point c). The increase of the external load S and the resulting change of the values of the internal forces at axial rigidity E_1A_1 and $E_1A_1^*$ versus the slope $w_{,x}(0)$ are shown in Fig. 3. In the bar of higher flexural stiffness the increase of the internal forces occurs but in the bar of lower flexural stiffness the internal force decreases.

In Fig. 4 the plot of bending moment at $x = 0$ and transverse deflection at $x = \frac{l}{2}$ versus force S is given. It should be noted here that only the moment M at the points of column supports tends to zero, but the bending moment $M_i(x)$ at $x \in (0, 1)$ tends to infinity.

The above diagrams are plotted for the physical and geometrical data given in the Table 1. The assumption of such data for calculation does not limit the generality of the obtained results.

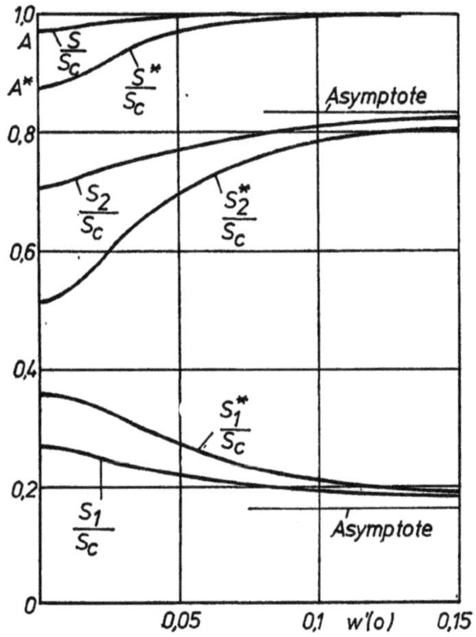

Fig. 3. External load S and internal forces S_1 and S_2 versus slope $w_{,x}(0)$

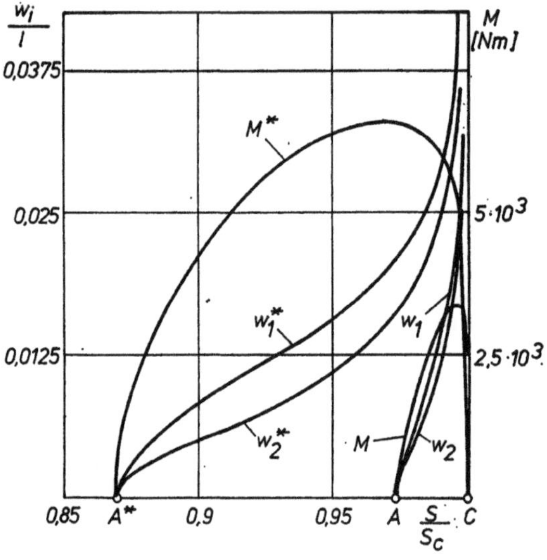

Fig. 4. Bending moment M at $x = 0$ and transverse deflection w_i at $x = l/2$ versus external load S

Table 1

Bar	E_i GN/m²	$I_i \cdot 10^6$ m⁴	$A_i \cdot 10^3$ m²	$A_i^* \cdot 10^3$ m²	$A_i^{**} \cdot 10^3$ m²	l m
1	2	3	4	5	6	7
1	100	2,0420	2,5133	5,0657	1,8786	4
2	200	5,2700	3,4557			4

References

1 SEVIN, E., On the elastic bending of columns due to dynamic axial forces including effects of axial inertia, J. Appl. Mech. Trans. ASME 27, 125–131 (1960).

2 KOUNADIS, A. N., Dynamic snap-through buckling of a Timoshenko two bar frame under a suddenly applied load, ZAMM 59, 523–531 (1979).

3 ARI-GUR, J.; WELLER, T.; SINGER, J., Experimental and theoretical studies of columns under axial impact, Int. J. Solids Structures 18, 619–641 (1982).

4 KOUNADIS, A. N.; GIRI, J.; SIMITSES, G. J., Nonlinear stability analysis of an eccentrically loaded two-bar frame, J. Appl. Mech. Trans. ASME 44, 701–706 (1977).

5 ROORDA, J., Stability of structures with small imperfections. J. Engng. Mech. Div. ASCE 91, EMI, 87–106 (1965).

Received May 9, 1983

Address: Doz. Dr. hab. LECH TOMSKI, Institute of Mechanics and Machine Design, Technical University of Częstochowa, ul. Deglera 35, 42–200 Częstochowa, Poland.

ZAMM · Z. Angew. Math. u. Mech. 65 (1985) 1, 61–62

GLICK, I. I.

More on Waiting Times

In the first part of this paper, by means of a Markov chain method similar to that used in [1], we obtain information about the waiting time for a particular sequence in sampling with replacement. Afterwards, it is shown how the results in [1] can be extended to a situation more general than that considered in [1].

Part I

Let $S = \{z_i\}_i$ be a set that contains at least two objects. S may be finite or countably infinite. Suppose we sample with replacement from S infinitely many times. Let the first selection be made at time 0, the second at time 1, etc. We then get a sequence, $(x_0, x_1, x_2, \ldots)$, of outcomes. We assume that the probability of selecting a particular member, z_i, of S at any given time is $\mathsf{P}(z_i)$, independent of the time of selection. Let $a_1 a_2 \ldots a_k$ be a prescribed sequence of length $k \geqq 2$ composed of elements of S. The sequence $a_1 a_2 \ldots a_k$ is arbitrary except that we assume that $a_i \neq a_1$ if $2 \leqq i \leqq k$. I.e., a_1 is not repeated. But there may be repetitions among $a_2, \ldots, a_k$. Let $p_i = \mathsf{P}(a_i)$ for $1 \leqq i \leqq k$. We consider the waiting time until the sequence $a_1 a_2 \ldots a_k$ first appears, and we derive, among other things, the pretty result that the expected waiting time until the first occurrence of $a_1 a_2 \ldots a_k$ is $(p_1 p_2 \ldots p_k)^{-1}$.

Let W denote the space of all sequences of outcomes $w = (x_0, x_1, x_2, \ldots)$. I.e., $W = S \times S \times \ldots$ Let W be supplied with the product measure. Define the random variables $X_0, X_1, X_2, \ldots$ on W as follows:

if $w = (x_0, x_1, x_2, \ldots)$, then

$$X_0(w) = \begin{cases} 1 & \text{if } x_0 = a_1 \\ 0 & \text{otherwise}, \end{cases}$$

and for $n \geqq 1$

$$X_n(w) = \begin{cases} 1 & \text{if } x_n = a_1 \\ j & \text{if } x_n = a_j \text{ and } X_{n-1}(w) = j-1, \text{ for } 2 \leqq j \leqq k. \\ 0 & \text{otherwise} \end{cases}$$

$\{X_n\}_{n=0}^{\infty}$ is a stationary Markov chain, and the state space is the finite set of integers $\{0, 1, \ldots, k\}$.

The transition matrix of the Markov chain is

$$\begin{pmatrix} 1-p_1 & p_1 & 0 & 0 & \ldots & 0 \\ 1-(p_1+p_2) & p_1 & p_2 & 0 & \ldots & 0 \\ 1-(p_1+p_3) & p_1 & 0 & p_3 & \ldots & 0 \\ \vdots & \vdots & \vdots & \vdots & & \vdots \\ 1-(p_1+p_k) & p_1 & 0 & 0 & \ldots & p_k \\ 1-p_1 & p_1 & 0 & 0 & \ldots & 0 \end{pmatrix}.$$

Let $p_{ij}^{(n)}$ denote the probability that starting from state i at some time the process is in state j n steps later. We adopt the usual conventions that $p_{ii}^{(0)} = 1$, and $p_{ij}^{(0)} = 0$ if $i \neq j$. Let $p = p_1 p_2 \ldots p_k$. Then, as can be calculated directly or read off from the powers of the transition matrix, we have

$$p_{0k}^{(n)} = \begin{cases} 0 & \text{if } 0 \leqq n \leqq k-1 \\ p & \text{if } n \geqq k, \end{cases}$$

$$p_{1k}^{(n)} = \begin{cases} 0 & \text{if } 0 \leqq n \leqq k-2 \\ p/p_1 & \text{if } n = k-1 \\ p & \text{if } n \geqq k, \end{cases}$$

$$p_{kk}^{(n)} = \begin{cases} 1 & \text{if } n = 0 \\ 0 & \text{if } 0 \leqq n \leqq k-1 \\ p & \text{if } n \geqq k. \end{cases}$$

Define T on W as follows: let $T(w)$ be the least integer n such that $X_n(w) = k$; let $T(w) = \infty$ if for every n it is true that $X_n(w) \neq k$. T is then the first time that the sequence $a_1 a_2 \ldots a_k$ occurs. Letting N denote the number of trials until the first time that the sequence $a_1 a_2 \ldots a_k$ occurs, i.e., N is the waiting time until the first occurrence of $a_1 a_2 \ldots a_k$, we have $N = 1 + T$.

For $n \geqq 1$ let $f_{0k}^{(n)}$ denote that conditional probability that starting from state 0 at some time the process enters state k for the first time n steps later, and let $f_{1k}^{(n)}$ denote the conditional probability that starting from state 1 at some time the process enters state k for the first time n steps later. Then

$$f_{0k}^{(n)} = \mathsf{P}(T = n \mid X_0 = 0), \quad \text{and} \quad f_{1k}^{(n)} = \mathsf{P}(T = n \mid X_0 = 1).$$

$$\mathsf{P}(\{X_0 = 0\} \cup \{X_0 = 1\}) = 1,$$

so

$$\begin{aligned} \mathsf{P}(T = n) &= \mathsf{P}(\{T = n\} \cap [\{X_0 = 0\} \cup \{X_0 = 1\}]) = \\ &= \mathsf{P}(\{T = n\} \cap \{X_0 = 0\}) + \mathsf{P}(\{T = n\} \cap \{X_0 = 1\}) \\ &= \mathsf{P}(T = n \mid X_0 = 0)\, \mathsf{P}(X_0 = 0) + \\ &\quad + \mathsf{P}(T = n \mid X_0 = 1)\, \mathsf{P}(X_0 = 1) = \\ &= (1 - p_1) f_{0k}^{(n)} + p_1 f_{1k}^{(n)}. \end{aligned}$$

As usual, $f_{0k}^{(0)}$ and $f_{1k}^{(0)}$ are defined to be 0. Let

$$P_{0k}(s) = \sum_{n=0}^{\infty} p_{0k}^{(n)} s^n, \quad P_{1k}(s) = \sum_{n=0}^{\infty} p_{1k}^{(n)} s^n, \quad P_{kk}(s) = \sum_{n=0}^{\infty} p_{kk}^{(n)} s^n,$$

$$F_{0k}(s) = \sum_{n=0}^{\infty} f_{0k}^{(n)} s^n, \qquad F_{1k}(s) = \sum_{n=0}^{\infty} f_{1k}^{(n)} s^n.$$

We have the well-known relations $P_{0k}(s) = F_{0k}(s) P_{kk}(s)$, and $P_{1k}(s) = F_{1k}(s) P_{kk}(s)$.

From the preceding expressions for the n-step transition probabilities,

$$P_{0k}(s) = \sum_{n=k}^{\infty} p s^n = \frac{p s^k}{1 - s},$$

$$P_{1k}(s) = \frac{p s^{k-1}}{p_1} + \sum_{n=k}^{\infty} p s^n = \frac{p s^{k-1}}{p_1} + \frac{p s^k}{1 - s},$$

$$P_{kk}(s) = 1 + \sum_{n=k}^{\infty} p s^n = 1 + \frac{p s^k}{1 - s} = 1 + P_{0k}(s).$$

Then

$$F_{0k}(s) = \frac{P_{0k}(s)}{P_{kk}(s)} = \frac{P_{0k}(s)}{1 + P_{0k}(s)} = 1 - \frac{1}{1 + P_{0k}(s)} = \frac{p s^k}{1 - s + p s^k},$$

$$F_{1k}(s) = \frac{P_{1k}(s)}{P_{kk}(s)} = \frac{P_{1k}(s)}{1 + P_{0k}(s)} = \frac{\dfrac{p s^{k-1}(1-s)}{p_1} + p s^k}{1 - s + p s^k}.$$

Note that $F_{0k}(1) = 1$ and $F_{1k}(1) = 1$. Then

$$\sum_{n=0}^{\infty} \mathsf{P}(T = n) = \sum_{n=0}^{\infty} [(1 - p_1) f_{0k}^{(n)} + p_1 f_{1k}^{(n)}] =$$

$$= (1 - p_1) \sum_{n=0}^{\infty} f_{0k}^{(n)} + p_1 \sum_{n=0}^{\infty} f_{1k}^{(n)} = (1 - p_1) F_{0k}(1) + p_1 F_{1k}(1) =$$

$$= 1 - p_1 + p_1 = 1 ,$$

so $\mathsf{P}(T = \infty) = 0$.

$$\sum_{n=1}^{\infty} p_{kk}^{(n)} = P_{kk}(1) - 1 = \infty ,$$

so state k is recurrent. States 0 and k communicate, and state 1 and k communicate, so

P (process enters state k infinitely often) $= 1$.

The generating function for T is

$$\mathsf{P}(s) = \sum_{n=0}^{\infty} \mathsf{P}(T = n)\, s^n = \sum_{n=0}^{\infty} [(1 - p_1) f_{0k}^{(n)} s^n + p_1 f_{1k}^{(n)} s^n] =$$

$$= (1 - p_1) F_{0k}(s) + p_1 F_{1k}(s) = \frac{p s^{k-1}}{1 - s + p s^k} .$$

Then $\mathsf{E}(T) = \mathsf{P}'(1) = (1/p) - 1$, so $\mathsf{E}(N) = 1 + \mathsf{E}(T) = 1/p$, the result promised earlier. (In particular, in the special case in which S consists of m equally likely elements, we get $\mathsf{E}(N) = m^k$.)

Let $f_{kk}^{(n)}$ denote the probability that starting from state k the first return to state k occurs n steps later. Let $F_{kk}(s) = \sum_{n=1}^{\infty} f_{kk}^{(n)} s^n$. Then, as is well-known,

$$F_{kk}(s) = 1 - \frac{1}{P_{kk}(s)} = 1 - \frac{1 - s}{1 - s + p s^k} = \frac{p s^k}{1 - s + p s^k} ,$$

so $F'_{kk}(1) = 1/p$.

This is the expected waiting time for the first return to state k starting from state k.

Hence, if $r \geqq 1$, the expected waiting time until the rth entrance into state k is $1/p + (r - 1)\, 1/p = r/p$.

Now we will calculate the variance of N. The variance of T may be expressed in terms of the first and second derivatives of $\mathsf{P}(s)$, but it is easier to use a different approach. Let $Q(s) = \frac{1 - \mathsf{P}(s)}{1 - s}$. Then, as is well-known,

$$\operatorname{Var}(T) = 2Q'(1) + Q(1) - Q^2(1) .$$

So, since

$$Q(s) = \frac{1 - \mathsf{P}(s)}{1 - s} = \frac{1 - p s^{k-1}}{1 - s + p s^k} ,$$

$$Q'(1) = 1 + \frac{1 - pk}{p^2} \quad \text{and} \quad Q(1) = \frac{1}{p} - 1 ,$$

and we have

$$\operatorname{Var}(T) = \frac{1}{p^2} + \frac{3}{p} - \frac{2k}{p}$$

$N = 1 + T$, so $\operatorname{Var}(N) = \operatorname{Var}(T)$.

(Again, if S consists of m equally likely elements, then $\operatorname{Var}(N) = m^k(m^k + 3 - 2k)$.)

Part II

We return briefly to the problem treated in [1], sampling from S until we get k successive repetitions of the same element of S. In [1] it was assumed that S consisted of $m \geqq 2$ equally likely elements, but now we allow S to be as in Part I of the present paper, i.e., the elements of S need not be equally likely, and S may even consist of countably infinitely many elements. We now let $p_i = \mathsf{P}(z_i)$, the probability of choosing the element z_i of S. Let $W = S \times S \times \ldots$ as above, and define the random variables $X_0, X_1, X_2, \ldots$ on W by

$X_n(w) =$ the number of consecutive occurrences up through time n of whatever the outcome at time n is,

as in [1]. The state space is the positive integers, and $\{X_n\}_{n=0}^{\infty}$ is a stationary Markov chain. The one-step transition probabilities are

$$p_{i1} = \mathsf{P}(X_{j+1} = 1 \mid X_j = i) = \mathsf{P}(X_{j+1} \neq X_j) =$$

$$= 1 - \mathsf{P}(X_{j+1} = X_j) = 1 - \sum_{r \geqq 1} p_r^2 \quad \text{for each } i \geqq 1 .$$

$$p_{i,i+1} = \mathsf{P}(X_{j+1} = X_j) = \sum_{r \geqq 1} p_r^2 ,$$

$$p_{ij} = 0 \quad \text{if} \quad j \neq 1 \quad \text{and} \quad j \neq i + 1 .$$

If we let $p = \sum_{r \geqq 1} p_r^2$, then the transition matrix becomes

$$\begin{bmatrix} 1 - p & p & 0 & 0 & \ldots \\ 1 - p & 0 & p & 0 & \ldots \\ 1 - p & 0 & 0 & p & \ldots \\ \cdot & \cdot & \cdot & \cdot & \cdot \\ \cdot & \cdot & \cdot & \cdot & \end{bmatrix},$$

just as in [1]. Hence, with the present meaning of p, the results of [1] are valid. In particular, the expected waiting time until the first time that we get k successive repetitions of the same outcome is $\frac{1 - p^k}{(1 - p)\, p^{k-1}}$.

After this paper was submitted for publication, the related paper [2] by G. Blom was brought to the attention of the author. Blom investigates the waiting time until the occurrence of an arbitrary, prescribed sequence. Blom, using a method different from that of the present paper, finds only the expected value of the waiting time and assumes that all the members of the set from which the selection is made are equally likely. Furthermore, it does not appear that Blom's method can be used if they are not equally likely. In the present paper, not only is the expected value of the waiting time obtained, but the probability generating function as well. In addition, the members of the set from which the selection is made are not assumed to be equally likely, and there may be infinitely many of them.

References

1 Glick, I. I., A note on runs, ZAMM 61 (1981), 119–120.

2 Blom, G., On the mean number of random digits until a given sequence occurs, J. Appl. Probab. 19 (1982), 136–143.

Received May 24, 1982, last revision November 30, 1983

Address: Dr. Irving I. Glick, Associate Professor, The George Washington University, Department of Mathematics, Washington, D.C. 20052, U.S.A.

ZAMM · Z. Angew. Math. u. Mech. 65 (1985) 1, 62–63

Hirche, J.

Some Remarks on Generalized Convexity of Sums and Products

In this note a few addenda to results of the recent paper by Schaible [5] and of one of his earlier papers [4] from 1977 are given. More general conditions are derived that exclude or ensure generalized concavity (convexity) of sums of absolute and relative terms, and of corresponding products, too.

Occupied with certain classes of composite functions, namely sums of absolute and relative terms as well as products, we study their quasiconcavity or explicit quasiconcavity (quasiconvexity). The notations used here are in accordance with Martos [3] and Hirche [2]. The property referred to by Schaible [5] as semistrictly quasiconcave is equivalent to explicitly quasiconcave for continuous functions.

Let us first consider the function

$$q(\boldsymbol{x}) = \boldsymbol{a}^{\mathsf{T}}\boldsymbol{x} + \boldsymbol{b}^{\mathsf{T}}\boldsymbol{x}/\boldsymbol{c}^{\mathsf{T}}\boldsymbol{x} \qquad (1)$$

on a convex set $S \subseteq R^n$. It is assumed that $\boldsymbol{c}^{\mathsf{T}}\boldsymbol{x} > 0$ for each $\boldsymbol{x} \in S$, $\boldsymbol{a} \neq \boldsymbol{0}$, $\boldsymbol{b} \neq \boldsymbol{0}$, and $\boldsymbol{b}, \boldsymbol{c}$ are linearly independent.

Distinguishing the cases

I. $\boldsymbol{a}, \boldsymbol{b}, \boldsymbol{c}$ are linearly independent;

II. $\boldsymbol{a} = \mu \boldsymbol{b}$, $\mu \neq 0$;

III. $\boldsymbol{a} = \lambda \boldsymbol{c}$, $\lambda \neq 0$,

Schaible [4] showed that in case I the function $q(\boldsymbol{x})$ is neither quasiconcave nor quasiconvex on an n-dimensional convex set S. Moreover, he gave conditions for the cases II and III which guarantee that $q(\boldsymbol{x})$ is explicitly quasiconcave or explicitly quasiconvex on S or on certain convex subsets of S, respectively. Herein the vector $\boldsymbol{a}$ is restricted to be an element of a one-dimensional subset of R^n.

Now let be

IV. $\boldsymbol{a} = \mu \boldsymbol{b} + \lambda \boldsymbol{c}$, $\mu \neq 0$ as well as $\lambda \neq 0$,

i.e. the vector $\boldsymbol{a}$ may be an element of a two-dimensional set. But, replacing $\boldsymbol{b}$ in (1) by the corresponding linear combination of $\boldsymbol{a}$ and $\boldsymbol{c}$ we get the case II with the addition of a constant (HIRCHE [2]). The constant does not affect the generalized concavity (convexity).

In the paper [2] it was proved that SCHAIBLE's statement for case I is also true if the condition

I′. $\boldsymbol{a}^T\boldsymbol{x}$, $\boldsymbol{b}^T\boldsymbol{x}$ and $\boldsymbol{c}^T\boldsymbol{x}$ are independent on S

is valid. The condition I′ is equivalent to the existence of such directions in S that the corresponding projections of $\boldsymbol{a}$, $\boldsymbol{b}$, $\boldsymbol{c}$ on these directions are linearly independent. Therefore, generalized convexity of the function $q(\boldsymbol{x})$ has to be excluded on certain sets S whose dimension is less than the number of the variables (e.g. equations in the definition of a polyhedral set S, in particular the feasible set of the transportation problem). Moreover, functions of the form (1) containing linear-affine functions instead of linear forms are included here (consideration in R^{n+1} subject to $x_{n+1} = 1$).

The functions investigated in the foregoing paper by SCHAIBLE [5] will be referred to as (S1) and (S2). We now consider the composite function

$$H(\boldsymbol{x}; \lambda, \mu) = \lambda h_1(\boldsymbol{x}) + \mu h_2(\boldsymbol{x}) + h_1(\boldsymbol{x})/h_2(\boldsymbol{x}) \tag{2}$$

on a convex set $S \subseteq R^n$ as a common generalization of (S1) and (S2). It is assumed that at least one of the functions $h_1(\boldsymbol{x})$ and $h_2(\boldsymbol{x})$ is nonlinear. Without loss of generality $h_2(\boldsymbol{x})$ is restricted to be positive on S. If $\mu = 0$ or $\lambda = 0$ we get (S1) and (S2), respectively. Hencefore let be $\lambda \neq 0$ as well as $\mu \neq 0$ in the following.

Giving the nominator of the quotient in (2) the form

$$(1/\lambda)\,[\lambda h_1(\boldsymbol{x}) + \mu h_2(\boldsymbol{x})] - (\mu/\lambda)\, h_2(\boldsymbol{x})$$

we get

$$H(\boldsymbol{x}; \lambda, \mu) = \lambda h_1(\boldsymbol{x}) + \mu h_2(\boldsymbol{x}) + {} + (1/\lambda)\,[\lambda h_1(\boldsymbol{x}) + \mu h_2(\boldsymbol{x})]/h_2(\boldsymbol{x}) - (\mu/\lambda)\,.$$

Since the constant summand $-\mu/\lambda$ does not affect the generalized convexity of $H(\boldsymbol{x}; \lambda, \mu)$ there is equivalence to (S1) where

$$f_1(\boldsymbol{x}) = h_1(\boldsymbol{x}) + (\mu/\lambda)\, h_2(\boldsymbol{x})\,, \qquad f_2(\boldsymbol{x}) = h_2(\boldsymbol{x})\,.$$

From the results in SCHAIBLE [5] we get conditions which ensure the generalized concavity or convexity of the function $H(\boldsymbol{x}; \lambda, \mu)$ and are summarized in the following two tables. Empty positions in the μ-column mean that μ is not restricted in sign in these cases. The abbreviation eqcx (eqcv) denotes *explicitly quasiconvex* (*explicitly quasiconcave*).

Table 1. ($\lambda > 0$, $h_2(\boldsymbol{x}) > 0$)

$h_1(\boldsymbol{x})$	$h_2(\boldsymbol{x})$	μ	$h_1(\boldsymbol{x}) + \frac{\mu}{\lambda} h_2(\boldsymbol{x})$	$H(\boldsymbol{x}; \lambda, \mu)$
convex	linear		$\geqq 0$	eqcx
	nonlinear	< 0		
concave	and concave	> 0	$\leqq 0$	eqcv
	linear			

Table 2 ($\lambda < 0$)

$h_1(\boldsymbol{x})$	$h_2(\boldsymbol{x})$		μ	$h_1(\boldsymbol{x}) + \frac{\mu}{\lambda} h_2(\boldsymbol{x})$	$H(\boldsymbol{x}; \lambda, \mu)$
concave	$\leqq -\frac{1}{\lambda}$	linear		$\geqq 0$	eqcv
		nonlinear	> 0		
convex		and convex	< 0	$\leqq 0$	eqcx
		linear			
concave	$\geqq -\frac{1}{\lambda}$			$\geqq 0$	
		nonlinear	< 0		
convex		and concave	> 0	$\leqq 0$	eqcv
		linear			

In a similar way as for (1) propositions on the generalized convexity of the function

$$p(\boldsymbol{x}) = (\boldsymbol{a}^T\boldsymbol{x} + \alpha)\,(\boldsymbol{b}^T\boldsymbol{x} + \beta)/(\boldsymbol{c}^T\boldsymbol{x} + \gamma) \tag{3}$$

may be obtained (HIRCHE [2]). It is natural to consider the composite function

$$P(\boldsymbol{x}) = [f(\boldsymbol{x}) + \alpha]\,[g(\boldsymbol{x}) + \beta]/[h(\boldsymbol{x}) + \gamma] \tag{4}$$

as a generalization of (3). In (4) let $f(\boldsymbol{x})$, $g(\boldsymbol{x})$ and $h(\boldsymbol{x})$ be not all linear and without loss of generality $h(\boldsymbol{x}) + \gamma > 0$ on S.

We distinguish the following cases:

A. The relation

$$(f(\boldsymbol{x}), \alpha) = \lambda(g(\boldsymbol{x}), \beta) + \mu(h(\boldsymbol{x}), \gamma)\,, \qquad \lambda \neq 0\,, \tag{5}$$

holds.

B., C., D. (5) does not hold, but

$$\begin{cases} f(\boldsymbol{x}) = \lambda g(\boldsymbol{x})\,, & \lambda \neq 0\,, \\ f(\boldsymbol{x}) = \mu h(\boldsymbol{x})\,, & \mu \neq 0\,, \\ h(\boldsymbol{x}) = \lambda f(\boldsymbol{x}) + \mu g(\boldsymbol{x})\,, & \lambda \neq 0 \text{ and } \mu \neq 0\,. \end{cases}$$

In case A we get

$$P(\boldsymbol{x}) = \lambda[g(\boldsymbol{x}) + \beta]^2/[h(\boldsymbol{x}) + \gamma] + \mu[g(\boldsymbol{x}) + \beta]\,.$$

Appropriate assumptions, as $\lambda > 0$, $h(\boldsymbol{x})$ concave, $g(\boldsymbol{x})$ linear or $g(\boldsymbol{x}) + \beta \geqq 0$ convex and $\mu > 0$, ensure that $P(\boldsymbol{x})$ is a convex (concave) function (BECTOR [1]).

In case B the function $P(\boldsymbol{x})$ can be represented by

$$P(\boldsymbol{x}) = [\lambda g^2(\boldsymbol{x}) + (\alpha + \lambda\beta)\, g(\boldsymbol{x}) + \alpha\beta]/[h(\boldsymbol{x}) + \gamma]\,.$$

It is assumed that the function $g(\boldsymbol{x})$ is linear or it has a range on which

$$\varphi(y) = \lambda y^2 + (\alpha + \lambda\beta)\, y + \alpha\beta$$

is a monotonic function. Then further suitable assumptions guarantee that the function $P(\boldsymbol{x})$ is explicitly quasiconcave (quasiconvex) on S or on certain subsets of S, respectively (MARTOS [3], Tables 3.1, 3.4, 3.5).

In case C the function $P(\boldsymbol{x})$ can be written as

$$P(\boldsymbol{x}) = \mu[g(\boldsymbol{x}) + \beta] + (\alpha - \mu\gamma)\,[g(\boldsymbol{x}) + \beta]/[h(\boldsymbol{x}) + \gamma]$$

and the violation of (5) implies $\alpha - \mu\gamma \neq 0$. The desired statements on $P(\boldsymbol{x})$ follow immediately from those given by SCHAIBLE for (S1). There is no necessity for listing the results here.

In case D we assume $f(\boldsymbol{x})$ and $g(\boldsymbol{x})$ to be concave and $f(\boldsymbol{x}) + \alpha > 0$, $g(\boldsymbol{x}) + \beta > 0$ on S. Then the function

$$1/P(\boldsymbol{x}) = \lambda/[g(\boldsymbol{x}) + \beta] + \mu/[f(\boldsymbol{x}) + \alpha] + {} + (\gamma - \lambda\alpha - \mu\beta)/[f(\boldsymbol{x}) + \alpha]\,[g(\boldsymbol{x}) + \beta]$$

is defined for each $\boldsymbol{x} \in S$. If the conditions $\lambda > 0$, $\mu > 0$ and $\delta = \gamma - \lambda\alpha - \mu\beta > 0$ are fulfilled then the function

$$\varphi(y_1, y_2) = \lambda/y_2 + \mu/y_1 + \delta/y_1 y_2$$

is for $y_1 > 0$, $y_2 > 0$ with respect to each of these variables strictly decreasing and besides convex (HIRCHE [2]). Hence it follows by means of well-known results (MARTOS [3], Tables 3.1 and 3.2) that the function $P(\boldsymbol{x})$ is explicitly quasiconcave on the assumptions introduced above. We do not intend to list here corresponding results belonging to other restrictions in sign and assumptions on the concavity or convexity of the functions $f(\boldsymbol{x})$ and $g(\boldsymbol{x})$.

References

1 BECTOR, C. R.: Programming problems with convex fractional functions, Operations Res. 16 (1968), 383–391.

2 HIRCHE, J., Verallgemeinerte Konvexität bei Summen und Produkten linearer und gebrochen linearer Funktionen, Wiss. Z. Univ. Halle, Math.-Nat. Reihe 32 (1983) 5, 91–99.

3 MARTOS, B., Nonlinear Programming. Theory and Methods, Akadémiai Kiadó, Budapest 1975.

4 SCHAIBLE, S., A note on the sum of a linear and linear-fractional function, Naval Res. Logist. Quart. 24 (1977), 691–693.

5 SCHAIBLE, S., Simultaneous optimization of absolute and relative terms, ZAMM 64 (1984) 8, 363–364.

Received May 7, 1983, revised version November 2, 1983

Address: Doz. Dr. JOACHIM HIRCHE, Martin-Luther-Universität, Sektion Mathematik, DDR-4020 Halle/S., Universitätsplatz 6

BOOK REVIEWS

Hafer, X. / Sachs, G., Senkrechtstarttechnik. Flugmechanik, Aerodynamik, Antriebssysteme. Berlin-Heidelberg-New York, Springer-Verlag 1982. XI, 406 S., 261 Abb., DM 82,—. US $ 38.20. ISBN 3-540-11075-5 (Hochschultext)

Im Vergleich zu den zahlreichen Lehrbüchern über die einzelnen Fachgebiete des Luftfahrtwesens, wie z. B. Aerodynamik, Flugmechanik oder Flugzeugstruktur, ist die Herausgabe eines Werkes über einen so speziellen Problemkreis wie die Senkrechtstarttechnik ein bemerkenswertes Ereignis. Das Ziel des Buches ist es, einen Überblick über die Entwicklung der Senkrechtstarttechnik und die verschiedenen technischen Lösungsmöglichkeiten zu geben, sowie die für den Entwurf von Senkrechtstartsystemen erforderlichen aerodynamischen und flugmechanischen Gesetzmäßigkeiten in Grundlagen und auch in den technischen Anwendungen darzustellen. Das vorliegende Buch ist der Behandlung von Flugsystemen vorbehalten, die die Fähigkeit zum Senkrechtstarten und -landen mit den Reiseflugleistungen konventioneller Flugzeuge verbinden. Deswegen wird die Problematik des Hubschraubers nicht mitbehandelt.

Das Buch beginnt mit einer Übersicht über die Entwicklung von Senkrechtstartsystemen, die alle bekannten, technisch interessanten Anordnungen von Senkrechtstartern in ihren wesentlichen Zügen mit durch Umrißzeichnungen ergänzten Merkmalen darstellt. Dabei werden die einzelnen Systeme kritisch bewertet. Danach folgt ein Kapitel, das den VTOL-Antriebssystemen gewidmet ist, in dem Strahl- und Raketentriebwerke sowie die Rotor- und Luftschraubenantriebe behandelt werden. Dabei werden sehr interessante Betrachtungen zur Triebwerksauswahl vom Standpunkt der verschiedenen Flugzustände angeboten. Weiter werden die Fragen der aerodynamischen Steuerung in verschiedenen Flugphasen, besonders vom Blickwinkel zusätzlicher Einrichtungen her analysiert, die Steuermomente um die Flugzeugfesten Achsen, sowie translatorische Bewegungen erzeugen können. Auch die Fragen der Flugmechanik der Senkrechtstarter werden gründlich besprochen. Dabei widmet man viel Aufmerksamkeit der Problematik des Schwebefluges und auch den damit eng verbundenen Fragen der Bodeneinflüsse und des Übergangsfluges. Für den in der Flugzeugbaupraxis stehenden Leser ist die ausführliche Erläuterung der Flugeigenschaftsrichtlinien und der -forderungen, die eng mit den bestehenden amerikanischen Spezifikationen MIL-I-83350 und AGARD Richtlinien Rep. Nr. 577 für V/STOL-Flugzeuge zusammenhängen, sehr wertvoll. Das abschließende Kapitel behandelt besondere Auslegungsprobleme und Einsatzbedingungen, die für die Entwurfsgestaltung und Einsatzmöglichkeiten von Senkrechtstartern sehr wichtig erscheinen.

Die Ausstattung des Buches ist, wie beim Springer-Verlag üblich, ausgezeichnet, obwohl es sich um einen Hochschultext handelt. Viele übersichtliche Abbildungen und Tafeln ermöglichen nicht nur eine effektive Einarbeitung in den Stoff, sie unterstützen weitgehend auch die Nutzung des Buches als Nachschlagewerk. Insgesamt stellt das Buch eine ausgezeichnete Kombination von Lehrbuch und Datensammlung dar, die in dieser Form auf dem Gebiet des Flugzeugentwurfs nur selten zu finden ist.

Prag — M. Píchal

NEW BOOKS

Gécseg, F. / Steinby, M., Tree Automata. Budapest, Akadémiai Kiadó, 1984. 235 S., Ft 295,—. ISBN 963 05 3170 4

Landau, L. D. / Liftschitz, E. M., Lehrbuch der Theoretischen Physik, Band X.: Physikalische Kinetik, Lifschitz, E. M. / Pitajewski, L. P., Berlin, Akademie-Verlag 1983. XIV, 480 S., M 35,—. BN 763 062 4 (5436/X)

Kavanau, J. L., Structural Equation Geometry. The Inherent Properties of Curves & Coordinate Systems. Los Angeles, Science Software Systems, Inc. 1983. 534 S., 61 Abb., $ 16.95. ISBN 0 937292-02-8

Augusti, G. / Borri, A. / Vannucchi, G. (eds.), Applications of Statistics and Probability in Soil and Structural Engineering. 4th International Conference, Università di Firenze, Italy, 1983. Bologna, Pitagora Editrice 1983. LIV, 1790 S., US $ 75.00. ISBN 88-371-0269-0

Hennig, K. (ed.), Random Vibrations and Reliability. Proceedings, IUTAM-Symposium Frankfurt/Oder (GDR) 1982. Berlin, Akademie-Verlag 1983. 367 S., M 68,—. BN 763 268 9 (6786)

Handbuch der Bedienungstheorie II. Formeln und andere Ergebnisse. Autorenkollektiv unter Leitung von Gnedenko, B. W. / König, D., Berlin, Akademie-Verlag 1984. 607 S., 13 Abb., M 98,—. BN 762 765 2 (6369/2)

Owen, D. R., A First Course in the Mathematical Foundations of Thermodynamics. Berlin-Heidelberg-New York-Tokyo, Springer-Verlag 1984. XVII, 134 S., 52 Abb., DM 74,—. US $ 27.60. ISBN 3-540-90897-8 (Undergraduate Texts in Mathematics)

Hale, J. K. / Magalhães, L. T. / Oliva, W. M., An Introduction to Infinite Dimensional Dynamical Systems — Geometric Theory. With an Appendix by Rybakowski, K. P., Berlin-Heidelberg-New York-Tokyo, Springer-Verlag 1984. VII, 195 S., 17 Abb., DM 54,—. US $ 20.20. ISBN 3-540-90931-1 (Applied Mathematical Sciences 47)

Ausloos, M. / Elliott, R. J. (eds.), Magnetic Phase Transitions. Proceedings of a Summer School at the Ettore Majorana Centre, Erice, Italy, 1983. Berlin-Heidelberg-New York-Tokyo, Springer-Verlag 1983. VII, 269 S., 103 Abb., DM 75,—. US $ 29.10. ISBN 3-540-12842-5 (Springer Series in Solid State Sciences 48)

Szilard, R. / Hartmann, F. / Ziesing, D., Finite Tragwerksanalyse mit dem Taschenrechner TI-59. Aufgaben- und Programmsammlung Berlin, Verlag Ernst & Sohn 1983. VIII, 268 S., 78 Abb., DM 66,—. ISBN 3-433-00981-3

Wilcox, C. H., Scattering Theory for Diffraction Gratings. Berlin-Heidelberg-New York-Tokyo, Springer-Verlag 1984. IX, 163 S., 4 Abb., DM 48,— US $ 17.90. ISBN 3-540-90924-9 (Applied Mathematical Sciences 46)

Fletcher, C. A. J., Computational Galerkin Methods. Berlin-Heidelberg-New York-Tokyo, Springer-Verlag 1984. XI, 309 S., DM 88,—. US $ 32.90. ISBN 3-540-12633-3 (Springer Series in Computational Physics)

Pironneau, O., Optimal Shape Design for Elliptic Systems. Berlin-Heidelberg-New York-Tokyo, Springer-Verlag 1984. XII, 168 S., DM 80,—. US $ 29.90. ISBN 3-540-12069-6 (Springer Series in Computational Physics)

Zeitschrift für Angewandte Mathematik und Mechanik

Volume 65 **1985** **Number 1**

Contents

Zeitschrift für Angewandte Mathematik und Mechanik is indexed in Current Contents/Physical, Chemical & Earth Sciences; Science Citation Index; Applied Mechanics Reviews; Mathematical Reviews, Zbl. Math.

Bei Fragen zur Produktsicherheit wenden Sie sich bitte an:
If you have any questions regarding product safety, please contact:

Walter de Gruyter GmbH
Genthiner Straße 13
10785 Berlin
productsafety@degruyterbrill.com